Alternative Project Delivery, Procurement, and Contracting Methods for Highways

SPONSORED BY
The Construction Institute (CI)
of the American Society of Civil Engineers

Construction Research Council

EDITED BY
Keith R. Molenaar, Ph.D.
Gerald Yakowenko, P.E.

ASCE

CI
Construction Institute

Published by the American Society of Civil Engineers

Library of Congress Cataloging-in-Publication Data

Alternative project delivery, procurement, and contracting methods for highways / sponsored by the Construction Institute (CI) of the American Society of Civil Engineers [and] Construction Research Council ; edited by Keith R. Molenaar, Gerald Yakowenko.

p. cm.

Includes bibliographic references and index.

ISBN-13: 978-0-7844-0886-5

ISBN-10: 0-7844-0886-6

1. Road construction contracts--United States. I. Molenaar, Keith R. (Keith Robert) II. Yakowenko, Gerald. III. Construction Institute. IV. Construction Research Council (Washington, D.C.)

TE195.A4355 2007

388.1'1--dc22 2006039310

American Society of Civil Engineers
1801 Alexander Bell Drive
Reston, Virginia, 20191-4400

www.pubs.asce.org

ISBN 13: 978-0-7844-0886-5
ISBN 10: 0-7844-0886-6
Manufactured in the United States of America.

Acknowledgments

The Construction Research Council (CRC) of the American Society of Engineer's (ASCE) Construction Institute (CI) has developed this special publication to provide a comprehensive and objective presentation of the use of alternative delivery, procurement, and contracting methods in the United States highway system. The CI was founded on the hypothesis that a need for cross cutting networking and technical exchange exists given the fragmented nature of the construction industry. The CI was established to provide a collaborative forum for every participant in the construction process and is organized into seven Directorates - Construction, Education and Research, Engineering, Equipment, Materials, Owners, and Services - providing an immediate focus area for every interest within the industry. The CRC is one of four committees under the Education and Research Directorate. The purpose of the CRC is to advance engineering knowledge and practice related to construction through stimulating and guiding research and assisting the financing thereof; to interpret and promote the utilization of the findings of research for the construction industry; and to identify future needs in the construction field and publicize them to stimulate appropriate research. The CRC has a membership of more than 100 of the premier construction engineering and management researchers and educators from around the world.

The CRC is endeavoring to develop special publications on various topics, project delivery included, for the benefit of ASCE membership and the construction industry that has so generously supported academic research and educational efforts. The CRC hopes that this effort will assist in advancing collaboration and research on project delivery methods. It is anticipated that the experiences, successes, and questions raised by the contributors to this publication will foster additional experimentation, research, and dialogue among all segments of the highway design and construction industry.

Early in this process of developing this publication, it was decided that this publication would contain rigorously peer-reviewed chapters. More than 15 CRC members were involved in the review process on a purely volunteer basis. The editors would like to sincerely thank all of the reviewers for their efforts and diligence in reviewing the papers in this publication. The editors would also like to thank the CI for their support in the production and dissemination of this publication. Without the CI and its membership, this work would not have been possible.

Introduction

The United States (US) highway system is the largest and most efficient network in the world. It was primarily built through a traditional design-bid-build delivery approach in which unit price construction contracts are awarded to the lowest bidder. The traditional approach was created to provide a transparent system of checks and balances between quality and cost. Federal and state highway agencies have the responsibility to design and deliver facilities that are safe and durable. Low bid procurement fosters an environment of fair and open competition that has allowed generations of workers to enter the market of public construction. Unit price contracts provide an equitable allocation of risks for quantities and price.

While the traditional project delivery approach has served the US public well, it has also received criticisms stemming from long delivery times, excessive cost growth and litigious relationships. Continuing to face increasing demands of the traveling public with declining staffs, federal, state and local agencies are employing alternative project delivery, procurement and contracting methods to improve the efficiency and effectiveness of public sector project delivery. In response to dissatisfaction by some stakeholders regarding cost, schedule, and quality performance, the Transportation Research Board (TRB) established a Task Force on Innovative Contracting Practices (A2T51) in 1987. This task force was created for the purpose of identifying promising innovative contracting practices for further evaluation. In December 1991, TRB published the final recommendations of Task Force A2T51 in a benchmark document entitled *Transportation Research Circular Number 386: Innovative Contracting Practices*. In 1990, the Federal Highway Administration (FHWA) implemented Special Experimental Projects 14 (SEP 14) to provide a means for evaluating some of the task force's more project-specific recommendations. While SEP 14 is still in use today to monitor innovative contracting methods, many innovative methods, such as time plus cost (A+B) bidding, lane rental, and warranties, have become mainstream and do not require SEP 14 approval on projects with federal aid in financing.

This publication provides a comprehensive and objective presentation of the use of alternative delivery, procurement and contracting methods in the US highway system. The following is a summary of articles included in this publication.

Project Delivery Approaches

"Key Implementation Issues and Lessons Learned with Design-Build Projects": The design-build project delivery method is a relatively new project delivery method for the highway industry. The successful implementation of this approach will ultimately depend upon the implementation of lessons learned by the State Departments of Transportation (DOTs). Gibson, O'Connor, Migliaccio, and Walewski evaluate the implementation of recommendations in the planning and procurement phases of the Texas DOT's $1.3 billion SH 130 design-build project. In order to gain the full benefits of design-build, the authors believe that Texas DOT and other contracting agencies need to understand, assess and allocate risks in a fair and cost effective manner. While the SH 130 is larger than most State DOT contracts, the lessons learned on SH 130 will serve as a benchmark for TxDOT and other State DOT interested in taking full advantage of this new project delivery method.

"Development of Performance Warranties for Performance-Based Road Maintenance Contracts": Another project delivery method that is seldom used in the US but has the potential for wider acceptance by the highway industry is the use of performance-based maintenance contracting. Ozbek and de la Garza explore the issues associated with the use of warranties in such contracts. The authors discuss the potential benefits and rationale for the use of warranties on performance-based maintenance contracts. The study includes a warranty clause template for the consideration of agencies interested in this approach.

"Miami Intermodal Center – Introducing "CM-At-Risk" to Transportation Construction": The "Construction Manager at Risk" project delivery method is an approach that is familiar to many in the vertical building construction industry; however, its use in the highway industry has been limited to a few non-traditional projects. Minchin, Thakkar, and Ellis discuss the potential benefits and issues associated with the use of CM-at-Risk for the first major transportation project in the US – the $1.35 billion Miami Intermodal Center – a large parking garage / car rental facility at Miami International Airport. The authors compare and contrast CM-at-Risk with other project delivery methods. They cite the following potential advantages in using CM-at-Risk: greater owner control of the design process than found in design-build, the ability to select a contractor with specialized expertise, cost control with a guaranteed maximum price, more effective use of constructability and value engineering expertise in the design phase of the project and more flexibility to deal with unforeseen changes in design. The authors noted, however, that the late acquisition of the construction manager seemed to cause issues and problems for one phase of the contract.

Procurement Methods

"Implementing Best-Value Procurement in Highway Construction Projects": Another alternate procurement method being evaluated by the State DOTs is best-value. This process provides for the use of both price and non-price based factors in the selection of the successful contractor. Gransberg, Molenaar, Scott and Smith analyze the procurement documents of over 50 best-value projects and present the results in a manner that will be helpful to State DOTs in developing procurement and contracting strategies to ensure quality and enhanced performance. The best-value selection plans are categorized into best-value parameters, evaluation criteria, scoring systems and award algorithms. The results of the project literature search, the case studies and a survey of contracting agencies indicate that there is a growing interest in the selection of a contractor with a proven record of success for special projects rather than relying on the traditional procurement system that is focused solely on the lowest initial price.

"Preference for A+B Contracting Technique Among State Departments of Transportation": One of the more commonly used alternate procurement methods by the State DOTs is the use of cost-plus-time bidding, or A+B bidding. Its recent popularity is based on the premise that is allows the owner to provide contractual incentives for early completion for projects that are subject to high road user costs. Strong, Raadt and Tometich performed a national survey of State DOT Construction Engineers and concluded that A+B bidding was one of the most effective contracting methods for each of the nine project types considered in the study. While the study confirmed the belief that A+B bidding shortens the project delivery time, it is important to note that the authors did not find evidence that A+B bidding increases internal administrative costs.

"Guidelines for Quality-Based Contractor Qualification": As State DOTs begin to consider options to the low-bid system of procurement, there has been increased interest in evaluating systems that incorporate quality factors in the qualification process. Minchin and Smith describe various quality-based performance measurement systems and provide a revised framework for the traditional contracting system used by State DOTs. The successful implementation of a quality-based qualification system would provide contracting industry with another incentive to integrate quality considerations in all phases of the construction operation so that they can remain competitive in an already competitive market.

Contracting Methods

"Guidelines for Warranty Contracting for Highway Construction": This paper received the best paper award for 2002 from the ASCE Journal of Management in Engineering and is being reprinted in this publication due to its significant impact on the industry. As the State DOT's financial and personnel resources continue to shrink in comparison with overall program needs, some owners have expressed increased interest in the use of warranties. Several State DOTs have evaluated pavement warranties and believe that this contracting technique will help them reduce administrative costs, reallocate performance risk, promote contractor innovation, increase the quality of the constructed product and ultimately reduce life cycle costs. Thompson, Anderson, Russell, and Hanna discuss guidelines for implementing warranty provisions for users that have little experience in this area. While the guidelines are intended for agencies that are considering the merits of a warranty program, they will also be helpful to contracting agencies with established warranty programs. The authors examine case study data from Wisconsin DOT's five-year asphalt pavement warranty program. This data shows a significant improvement in the quality of construction when comparing ride and distress values for warranted and non-warranted pavement sections.

"Incentive / Disincentive Contracting Practices for Transportation Projects": State DOTs have found contractual incentives / disincentives (I/D) provisions to be very effective in achieving higher quality and early completion. In particular, the use of I/D provisions and the cost-plus-time bidding technique have generally resulted in great success in reducing the construction time to deliver a critical phase or phases of a project. Sillars and Leray review and consolidate many sources of information regarding I/D provisions for early completion. The authors provide an overview of concepts associated with the successful use of I/D provisions and a discussion of the implementation process that should be used. This process includes the identification of project goals, the selection of potential candidate projects, incentive types, risk management, preparation of specifications, contract administration and an evaluation of the process. The paper includes a model specification for the use of cost-plus-time bidding with and I/D provision.

Contents

Key Implementation Issues and Lessons Learned with Design-Build Projects 1
G. Edward Gibson, Jr., James T. O'Connor, Giovanni Migliaccio, and John Walewski

Development of Performance Warranties for Performance-Based Road Maintenance Contracts 20
Mehmet Egemen Ozbek and Jesús M. de la Garza

Miami Intermodal Center—Introducing "CM-at-Risk" to Transportation Construction 46
R. Edward Minchin, Jr., Ketan Thakkar, and Ralph D. Ellis, Jr.

Implementing Best-Value Procurement in Highway Construction Projects 60
Douglas D. Gransberg, Keith R. Molenaar, Sidney Scott, and Nancy Smith

Preference for A + B Contracting Technique among State Departments of Transportation 80
Kelly C. Strong, Nolan Raadt, and James Tometich

Guidelines for Quality-Based Contractor Qualification 98
R. Edward Minchin, Jr. and Gary R. Smith

Guidelines for Warranty Contracting for Highway Construction 113
Benjamin P. Thompson, Stuart D. Anderson, Jeffrey S. Russell, and Awad S. Hanna
From *J. Manage. Eng.* Vol. 18, No. 3, July 2002, pp. 129–137. Reproduced with permission.

Incentive/Disincentive Contracting Practices for Transportation Projects 129
David N. Sillars and Jean Pol Armijos Leray

Index 153

Key Implementation Issues and Lessons Learned with Design-Build Projects

G. Edward Gibson, Jr.[1], James T. O'Connor[2], Giovanni Migliaccio[3] and John Walewski[4]

Abstract

Public agencies across the country are pursuing the design-build (DB) project delivery method to improve cycle-time performance on projects, and numerous transportation departments are currently using DB for highway construction. This paper outlines the findings of two studies of design-build projects for the Texas Department of Transportation (TxDOT). The first of these studies included the development of a DB implementation guide for TxDOT and related key implementation issues needed to perform DB on transportation projects. The second involves an on-going investigation into lessons-learned during planning, procurement and execution phases of the State Highway 130 (SH130) project currently underway in Central Texas. The SH130 project is a 49 mile-long toll road being constructed under a Comprehensive Development Agreement (CDA). The project includes design, construction, and maintenance of a four-lane expressway within the specified corridor with a cost of approximately $1.3 billion and a scheduled completion date in 2007—the first DB project and the largest single highway project undertaken by TxDOT. This paper outlines critical success factors of design-build transportation projects for TxDOT, as well as early lessons-learned on the SH130 project with particular attention given to enabling management issues, contract management, right of way and utility relocation issues, and critical relationship factors.

Introduction

Highways play a large role in the economic development of many states and as infrastructure needs continue to grow, benefits can be achieved from techniques that improve the speed and effectiveness of project delivery. While other public entities are utilizing alternative project delivery methods and contracting approaches with increasing frequency, Texas state law has until recently limited the Texas Department of Transportation (TxDOT) to the design-biDBuild (DBB) project delivery method

[1] Professor and Garry Neil Drummond Endowed Chair in Engineering, University of Alabama, Department of Civil, Construction and Environmental Engineering, 259 H. M. Comer, Box 870205, Tuscaloosa, AL, 35487-0205, 205-348-6987, fax: 205-348-0783, egibson@eng.ua.edu.

[2] C. T. Wells Professor of Project Management and Civil Engineering, The University of Texas at Austin, Department of Civil Engineering, 1 University Station C1752, Austin, TX 78712-0276.

[3] Graduate Research Assistant, The University of Texas at Austin, Department of Civil Engineering, 1 University Station C1752, Austin, TX 78712-0276.

[4] Graduate Research Assistant, The University of Texas at Austin, Department of Civil Engineering, 1 University Station C1752, Austin, TX 78712-0276.

along with a few innovative contracting approaches such as lane rental, partnering, and A+B contracting.

The design-build delivery method was recently introduced in Texas with State legislation that allows TxDOT to adopt design-build and other innovative processes by entering in a "Comprehensive Development Agreement" (CDA) that facilitates a public-private partnership. The CDA is currently the statutory approach for adopting innovative project delivery methods in the State of Texas and the Texas Transportation Code outlines the boundaries for a CDA as:

> *An agreement with a private entity that, at a minimum, provides for the design and construction of a transportation project and may also provide for the financing, acquisition, maintenance, or operation of a transportation project (Texas Transportation Code, Title 6, Section 370.305, subsection (b)).*

Although unique to Texas in many ways, this approach agrees with the U.S. Code definition of design-build contract as:

> *an agreement that provides for design and construction of a project by a contractor, regardless of whether the agreement is in the form of a design-build contract, a franchise agreement, or any other form of contract approved by the Secretary (of Transportation)* (U.S. Code Title 23, Section 112).

This paper outlines the findings of two related studies regarding design-build for TxDOT. The first of these studies included the development of a design-build (DB) implementation guide for TxDOT and related key implementation issues needed to perform DB on transportation projects (Gibson and Walewski 2001, 2002, and Walewski, Gibson, and Jasper 2001). The second involves an on-going investigation into lessons-learned during the planning, procurement and execution phases of the CDA-related State Highway 130 (SH130) and the State Highway 45 South-East (SH45 SE) projects currently underway in Central Texas (O'Connor, et. al. 2004a, 2004b, 2004c).

As these research projects proceeded, the research team performed an extensive literature review, conducted numerous interviews as well as source document reviews, and attended six industry forums and conferences to determine current practices related to project delivery methods and contracting approaches. A detailed review of the legal and regulatory status was performed to determine which procurement tools were legal and available for use by TxDOT at the time of the study. In the first study, 13 structured interviews were conducted with experts from the construction industry, the legal community and academia for the research effort to assess the status and availability of delivery methods and contracting approaches. Conversations were held with an additional 18 industry and academia contacts to solicit specific information. Four meetings were held with the TxDOT project team to provide input and guidance to the research team's work and direction. As part of the second research project, researchers conducted 19 interviews with SH 130 participants from the owner and contractor perspective, and attended 2 industry conferences, and an open forum public meeting. Researchers received input and guidance from TxDOT project team on a regular basis through progress meetings, phone calls and e-mail. The lessons learned given later in this paper were vetted by the research sponsors.

For the purposes of this paper, a project delivery method equates to a procurement approach that defines the relationships, roles, and responsibilities of project team members and the sequence of activities required to complete a project. An innovative contracting approach is a specific procedure used under the larger umbrella of a procurement method to target specific activities or objectives of a project.

Project Delivery Methods and Innovative Contracting Approaches

The traditional project delivery method for highway construction projects in Texas has been DBB where design and construction occur in sequence with solicitation steps for each. Using this method, TxDOT develops the project scope, then contracts with design and engineering firms and once the designs and specifications are completed, bid solicitation follows and then the construction contract is awarded to the qualified bidder with the lowest price. Because the steps are followed sequentially, firm costs can usually be established and the design and contractor selection is simplified. DBB provides minimal interaction between the designer and contractor, generally omits reviews for constructability cost savings, and often creates an adversarial relationship between the parties.

Outside the use of DB under the provisions of the CDA legislation that allows TxDOT, and Regional Mobility Authorities to use this project delivery method, authorities are still limited to the traditional DBB project delivery method and competitive bidding is required for highway improvement contracts. Under the provisions of the CDA legislation, the prescribed entities may solicit proposals or accept unsolicited proposals and final contractor selection is based on the concept of "best value".

In the review of procurement practices by other public entities, the authors found a proliferation of construction project delivery methods and contracting approaches as well as considerable confusion on their application and use. In response, a matrix of the most relevant project delivery methods for highway projects and their various attributes was developed for TxDOT (Appendix A). The matrix allows for a simplified cross-comparison of the pros and cons of each method, responsibilities of the parties involved, general assumptions concerning which projects each is best and least suited for, and generalizations on how each method impacts quality, schedule, cost-control, and legal liability. The matrix is intended to provide TxDOT with a simple overview analysis of which delivery method is applicable for a specific project.

Most state DOTs are just beginning to venture past contracting approaches that dictate exact methods and prescriptive specifications on how the work is to be done (Hancher 1999). In recent years, state DOTs have increasingly used contracting strategies to supplement project delivery methods for added long-term benefits and contracting approaches such as constructability reviews, incentives/disincentives, performance-related specifications, transfer of quality control, and warranties are available regardless of the project delivery method used (Loulakis and Huffman 2000). As previously noted, TxDOT has combined contracting approaches such as A + B contracting, partnering, and lane rental with DBB on a limited basis. The initial research effort documented many of these methods and suggested that TxDOT should build upon the work of the Transportation Research Board National Cooperative Highway Research Program (TRB 1991, 1996, 1999) and others (Federal Highway Administration 1998, American Association of State Highway and Transportation Officials 1998) to develop comprehensive guidelines for implementing non-traditional contracting approaches for highway construction.

Selecting the appropriate methods and approaches requires an adequate understanding of each delivery method and contracting approach and the ability to distinguish their differences. By providing a method of comparing the pros and cons of the different project delivery methods and descriptions of the various contracting approaches, the potential for good management decisions on their use can be enhanced. The selection of project delivery methods and contracting approaches should be a project-specific decision *process* and is very dependent on the type of project, project risk profile, human resources available, and specific objectives of the project.

The Design-Build Option

In Texas, alternative delivery methods such as DB and CM at risk have been legislatively authorized for school districts and institutions of higher education for buildings and other facilities, and this authority has also been granted to the Texas General Service Commission, as well as cities and counties (AGC-Texas 1998, Gallegos 2001, Texas Controller 2000). As addressed, by specific legislation beginning in 2001, TxDOT, the Texas Turnpike Authority, and the newly created Regional Mobility Authorities gained the ability to use DB with the framework of a CDA. TxDOT (Pensock 2004) perceives that a CDA:

- Accelerates project through concurrent activities
- Fixes price early in design phase
- Guarantees completion date
- Designates single point of responsibility
- Allocates project risks among parties
- Provides life cycle cost efficiency
- Requires continuous constructability reviews
- Streamlines project administration
- Enhances cost control and risk management

Existing CDA contracts outline the approach for Major Subcontracts (in excess of $3,000,000) by the Design-Builder. In fact, the latest CDA contract for SH45 SE allows the DB contractor to "submit alternative procedures to TxDOT for approval" (TxDOT 2004, subsection 7.3.4, pp.41). In this way, the application of innovative contracting methods, such as the ones outlined in Appendix A, is allowed. Alternatively, the first contract for SH 130 prescribes that subcontracts will go to "lowest responsive bid by a responsible bidder" (TxDOT 2001, subsection 19.3.4, pp.118).

Based on the literature review and interviews conducted for this study, the primary reason DB contracting is selected by public and private owners is to shorten the duration on specific projects by melding the design and construction processes. Quality, cost-effectiveness, and a single point of responsibility are also cited as reasons to pursue DB. Furthermore, DB can allow owners to establish cost and schedule, promote innovation, reduce claims, and foster a team approach that encourages communication. Early collaboration on projects between designers and contractors usually enhances their relationship, and often results in change orders avoidance because the process encourages the contractor to point out problems in the design or constructability issues early in the bidding process. The owner's dilemma to determine whether the contractor or the designer is at fault for changes is reduced when a single source of responsibility exists.

However, DB can limit competition during the bidding process. With DB, an owner puts the project out to bid and design-builders may be reluctant to develop proposals without the benefit of complete plans. Comparison of projects proposals can be difficult because each of the proposers is responding to limited guidance and final solutions can vary widely. Problems also arise when an owner has an ill-prepared project and equally ill-defined DB selection criteria.

Determining when it is appropriate to use DB, and on what type of projects, are critical steps in gaining the advantages that the process can provide. Projects where speed is desirable to reduce the impact to the traveling public, or when innovation is desired from the design team are often good candidates for DB.

For TxDOT, the application of DB via the CDA process should be seen as a precursor step to develop a comprehensive DB pilot program. In the initial work for TxDOT, the authors advised that TxDOT should also develop assessment criteria for selecting candidate DB projects because forcing the wrong project into a DB contract may diminish or eliminate potential benefits. A guidebook with example guidelines, procedures, and process maps was developed by the authors to assist TxDOT in the transition to achieve proficiency with the DB project delivery system (Gibson and Walewski 2001). The current research project investigating lessons learned from the SH 130 project is also intended to develop a benchmarking process for future DB projects.

Design-build has the potential to benefit TxDOT as an alternative form of delivering highway construction projects and a supplement to DBB. Nevertheless, for TxDOT to gain the full benefits of DB, it needs to understand, assess, and allocate the associated risks as well as determine a process to implement the methodology. In summary, in the first study the authors believed that TxDOT must:

1. *Develop DB process guidelines and a delivery process (planning, scope, RFP, selection, management, etc.).* DB is a unique, distinct project delivery method so the associated guidance documents should be developed specifically for this procurement method.
2. *Assess the availability of the skills required for the use of DB in the organization.* Experience with DB contracting enhances the chances for success and limits the risk to the parties involved. If TxDOT lacks the necessary skills and experience to undertake DB, consideration should be given to obtaining professional services from an experienced firm to assist with preparing the necessary documents and performing the required tasks.
3. *Train selected members of the organization in the use of the DB project delivery system.* DB contracting requires a different skill set than administrating traditional DBB contracts for highway construction. To

perform these tasks adequately, TxDOT staff involved with DB project delivery should receive adequate training to gain the required knowledge needed.

4. *Optimize communication among the parties involved within TxDOT.* DB projects require more project coordination at the onset of the project planning phase and will require the design and construction divisions of TxDOT to integrate and coordinate on a much grander scale than currently exists.
5. *Optimize the pre-project planning process.* TxDOT must develop the skills to create a detailed scope package for DB and develop reasonable submission requirements. Overly detailed RFP proposals may reflect a lack of understanding of the project scope and can be financially burdensome for the bidders as well as TxDOT. Proposals should be limited to the information necessary to make judgment based on the merits of the proposals.
6. *Select pilot DB projects that have a relatively certain scope and contain well-known processes and technologies.* Although DB can be used on all types of highway-related construction, TxDOT should select projects with which it has adequate experience for the initial phase of the pilot program.
7. *Ensure selection of qualified DB contractors.* Prequalification of contractors should limit the final competitors to those with adequate experience and financial resources. A balanced evaluation process should be administered by individuals who understand the design and construction constraints specific to the project.
8. *Develop succinct criteria specifications.* The project requirements listed in the RFP should be designed in performance terms rather than a more prescriptive manner that may limit creative solutions.
9. *Develop a systematic way to evaluate project results to determine if existing DB procedures and approval processes are adequate, and respond to legislative requirements.*

It should be noted that these same steps should be followed by any organization venturing into DB for the first time. How these recommendations have been addressed by TxDOT will be discussed in the conclusions of this paper.

State Highway 130 Project Overview

Rapid population growth and commercial growth in and around Austin, Texas, combined with an inadequate transportation network has contributed significantly to its ranking as having the worst traffic delays among medium size cities in the Nation (Schrank and Lomax 2004). In order to solve this problem, the Department of Transportation of Texas (TxDOT) is implementing the Central Texas Turnpike Project (CTTP) with a funding amount of about $3.6 billion (http://www.texastollways.com/tta/investors_schedule.asp). As shown in Figure 1, the turnpike will be constituted by five toll ways: SH45 N, Loop 1 Extension, US183A, SH130, and SH45 SE. The system will create a bypass to Interstate 35 with southern connector on I-35 (SH45 SE) and northern connector on I-35, Mopac-Loop 1 and US183 (SH45 N). A diagram of the entire CTTP is given in Figure 1.

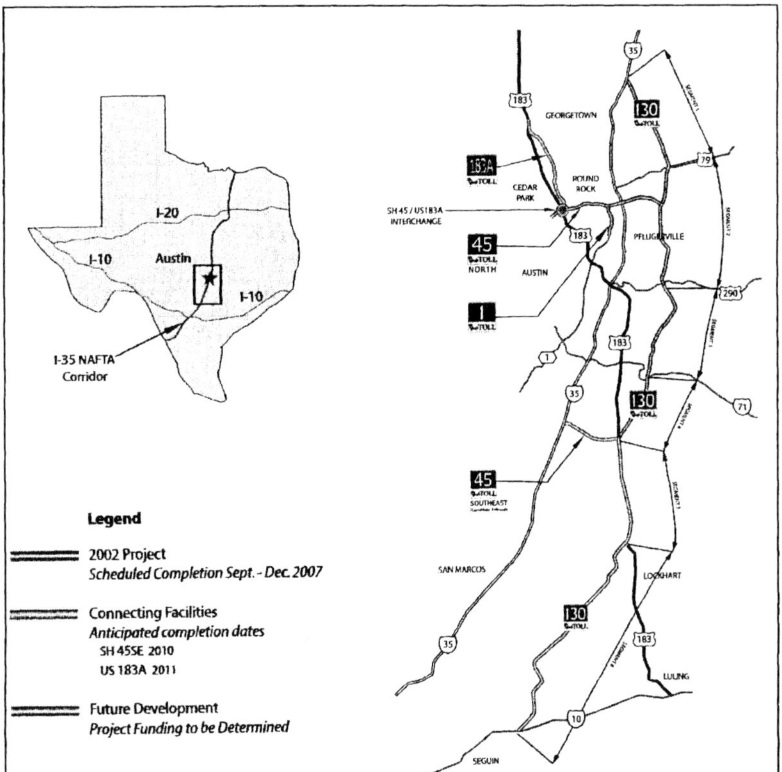

Figure 1: Central Texas Turnpike System (adapted from TxDOT website: http://www.texastollways.com/tta/maps/system_map.pdf)

In 2002, TxDOT selected as design-builder for the SH130 project, Lone Star Infrastructure (LSI), a joint-venture created specifically for this project. The parties entered into a CDA for a total amount of $1.3 billion that is the largest contract awarded by TxDOT in its history. At completion, the project will include six segments for a total of 91 miles from Interstate Highway 35 (IH-35) at State Highway 195 (SH 195), north of Georgetown, Texas, to Interstate Highway 10 (IH-10), near Seguin, Texas, and will be a four-lane divided facility with major interchanges at SH 195, IH-35, SH 29, U.S. 79, SH 45 North, U.S. 290 and SH 71. The Department reserves the option to extend the highway to I-10 near Seguin by completing the remaining two segments, if future funding becomes available. The contract also has an option that LSI will provide maintenance for an initial term with the opportunity of two extensions. The maximum term of the Maintenance Agreement, including both extensions, is 15 years.

Table 1: SH 130 Project Status as of March 2005.

ROW		Total number of parcels	408
		Parcels acquired for TxDOT through other entities (e.g. , Williamson County)	12 (3%)
		Number of parcels to be acquired by LSI	396 (97%)
		Parcels with first offer made	343 (84%)
		Parcels acquired by negotiation	144 (35%)
		Parcels acquired by condemnation	74 (18%)
		Total number of parcels acquired	230 (56%)
		Parcels not acquired with possession and use agreements	44 (11%)
		Parcels available for construction	274 (67%)
Utilities		Total number of Utilities	437
		Total to adjust	310
		Total number of required utility assemblies	130
		Assemblies in revision	28 (21%)
		Assemblies approved for construction	60 (46%)
Design	Roadway	Segment 1 (% of complete)	99%
		Segment 2 (% of complete)	99%
		Segment 3 (% of complete)	93%
		Segment 4 (% of complete)	69%
	Bridge	Total Number of Bridges	123
		100% Plans Review Complete	88 (71%)
		% Plans Certified	71 (58%)
Construction		Segment 1	Ongoing
		Segment 2	Ongoing
		Segment 3	Commenced
		Segment 4	Commence in December 2005
Contract Status		Original contract amount	$1,306,554,920
		Authorized amount	$998,955,914
		Authorized Change Orders (COs)	8
		Authorized COs Amount	$ 52,535,479
		Current authorized Contract Amount	$1,051,491,393
		Approved Payments	$380,473,148
		Amount remaining	$671,018,245
		Percent Dollars Expended	36.2%
Schedule		Authorized Time Adjustments	None
		Evaluation of Critical Path	On schedule

Most of the project financing relies on the issuance of revenue bonds. In order to create a favorable financing scenario, it was fundamental to decrease any schedule uncertainty by assuring guaranteed completion dates. Therefore, the Department decided to deliver the turnpike backbone, SH130, and its southern connector to I-35 (SH45SE) via the use of the CDA approach. With this approach, project tasks are distributed between the Department and the design-builder. In that way, TxDOT can anticipate earlier design and construction. This approach will allow TxDOT to build these two projects in a shorter time frame. As a result, the Department is planning to use a similar approach through the State for other high-priority projects like the proposed $150 billion Trans Texas Corridor.

In 2002 TxDOT issued LSI notice to proceed for design, procurement of Right-Of-Way (ROW), utility adjustment services and construction on segments 1 to 4 (subdivided in 15 sections) for a total of 49-miles. The current status of the project is given in Table 1.

Lessons Learned to Date

Lessons learned reported in this section relate to the planning, procurement and execution phases of the SH130 project, but also incorporate lessons learned during the procurement of the SH45 SE project that followed SH130.

Project Planning

- *In organizing the project team:*
  - *Consider the use of external consultants.* If the Department is not familiar with DB, the use of external consultants on the owner's team is beneficial because it incorporates the experiences acquired from other DB projects and/or other states.
  - *Address team leader expertise in managing complex communication environments.* The extensive number of entities in DB projects, including multi-party contractors, public interest groups, and political figures, requires a team leader with outstanding communication skills.
- *Other planning issues:*
  - *Track personal characteristics of new and existing personnel.* The needed characteristics for TxDOT personnel on CDA projects are flexibility, pace, and accuracy. Recruitment of personnel with these characteristics requires time, so HR managers should investigate and track employees and perspective employees with these characteristics on a regular basis.
  - *Consider co-location of highways and railroad in a timely fashion.* Railroads have an acceptable slope that is different from highway projects (1% vs. 3%). Consequently, this difference will affect the footprint and the cost of the facility. Therefore, timing is important to set corridor boundaries and to activate adequate financing flow. The political and legislature support for a railway in the corridor needs to be identified early in the planning process. For SH 130, the railway requirements were identified too late (after contract award) to be included in the project.

Project Procurement

The SH130 and SH45 SE projects were awarded following a two-step evaluation process that included a prequalification/short-listing phase and a proposal evaluation phase. This type of selection is usually adopted for projects that are more complex in order to contain costs of proposal evaluation by short-listing interested parties based on qualification before assessing the evaluation of the proposals. Figure 2 represents an overview of the procurement process at phase/sub-phase level with recommended durations at sub-phase level and milestones adopted for the breakdown (O'Connor et al., 2004a).

Initially, the Department prepared a Request for Proposals and Qualifications (RFPQ), interacted with interested parties for facilitating their submittals and finally evaluated the Proposals and Qualifications Submittal (PQS) before it released a shortlist of qualified proposers.

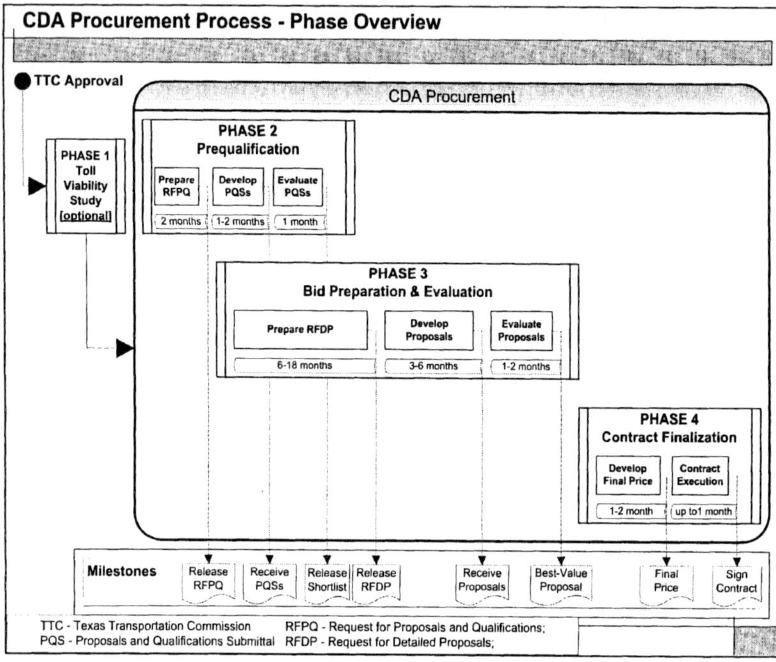

Figure 2: Overview of Comprehensive Development Agreement Procurement Process with Schedule and Milestones (Source: Migliaccio, Gibson, and O'Connor 2005)

During the second stage, the Department prepared a Request for Detailed Proposals (RFDP). This document passed through an interactive stage with the short-listed firms and risks were allocated between project parties based in part on these discussions. These meetings were held in accordance with FHWA guidelines and any clarifications were provided through addenda sent to all short-listed firms. After the public release of the final RFDP to the short-listed firms, the Department interacted with the interested parties and facilitated their proposal submittals by scheduling recurring rounds of one-on-one meetings. Finally, an evaluation of the submitted proposals was conducted to determine the proposal offering the best-value to the State of Texas and to put forward a recommendation to the State Transportation Commission

Because the initial planning phase (Phase I) of the project were outside the boundaries of this research study, the lessons learned concerning this part of the procurement process are not a part of this analysis.

PHASE II: Prequalification

- Before start preparing the RFPQ:
    - Prepare a standardized RFPQ document and make it and RFPQ documentation from other CDA projects available for consultation to the project team. The DB delivery approach requires different RFPQ documentation than traditional projects. Employees need to become familiar with this new documentation in order to provide quick and effective feedback to legal consultants. The availability of a standardized documentation also allows other individuals in the department to know how the RFPQ will look and, ultimately improve their understanding of its preparation. Standardized RFPQ documents will streamline this phase of the procurement process and will decrease the duration of the review activity.
    - Achieve a basic understanding of the project description in terms of location, characteristics, scope of work, and risk allocation. A clear identification of the needed input for the RFPQ document combined with a standardization of the document itself will help shorten the preparation of the RFPQ document.

- Early in the process of preparing the RFPQ document, release a status report to legal counselors outlining the project's development, environmental clearance process, and amount of preliminary engineering to be included in the RFPQ. Legal counselors need to be aware of the status of concurrent activities in order to develop a good RFPQ draft and to understand what is going to be included in the document.
- In preparing the PQS Evaluation process, develop a suggested evaluation schedule before appointing the evaluation subcommittee members. Understaffing a subcommittee can slow down the evaluation dramatically because the entire process has to wait as the understaffed subcommittee completes its evaluation. A defined schedule helps the process manager to understand the needed size and qualifications to achieve a streamlined evaluation process. This lesson is also applicable to the proposals evaluation process (Prepare RFDP phase).
- In interacting with interested parties for developing PQSs, have one-on-one meetings with interested firms if the project includes bonds or developer financing options. Direct interaction with proposers allows the DOT team to probe the reactions of the interested parties in terms of the requirements, and to

take any necessary corrective action. Again, these meetings should take place in accordance with applicable regulations in order to ensure fairness.

PHASE III: Bid Preparation and Evaluation

- *In developing the RFDP document:*
    - *Start developing technical attachments earlier in the process to decrease process duration.* Development of technical attachments is a critical activity for defining details of the agreement and for releasing the RFDP document. Surveying and mapping activities should begin as soon as possible.
    - *Conduct interactive sessions between attorneys, engineering consultants, and DOT employees early in the development of the RFDP.* These meetings will improve the attorneys' understanding of what is entailed in the technical provisions, and decrease the risk of overlapping or missing information in the contract.
- *In conducting the industry review phase:*
    - *Identify and monitor the status of critical path activities.* While a trade-off exists between project schedule and the necessity to conduct adequate industry reviews, it is common to extend this phase to allow for the completion of other mandatory activities. Environmental clearance and preliminary engineering studies are two of the activities that should be closely monitored.
    - *Establish the number of one-on-one meetings with short-listed bidders depending on project complexity, and procurement schedule pressure.* The number of necessary meetings depends mostly on project complexity. If the procurement schedule necessitates an early release of the RFDP document, be aware that more time will be needed for interaction during proposal development and more addenda will be needed.
- *In interacting with short-listed firms for developing detailed proposals:*
    - *Allocate sufficient time between issuing the RFDP and the first round of meetings.* Proposers need adequate time to thoroughly analyze the documents before interacting and making comments.
    - *Schedule two different rounds of one-on-one meetings with a sufficient time between each.* Usually, addenda are released after each round of meetings, so the time allotted should be adequate for the legal counsel to revise the document, distribute addenda, and have the proposers analyze it.

- *In interacting with short-listed firms for selecting Alternative Technical Concepts (ATC):* Limit the number of ATCs that each proposer can submit. Evaluating ATCs is time-consuming, and can slow down the process. Therefore, there is a need for a pre-screening process to limit the effort in evaluating ATCs. For instance, defining a minimum dollar amount threshold for cost-saving ATCs can avoid time-consuming evaluations on less cost effective ATCs. These less cost effective ideas can addressed in the design or construction phases as Value Engineering proposals.

Project Execution

Initial lessons learned in the project execution phase, regarding project management, contract management, right-of-way and utility relocation issues, include:

- <u>Project management</u>
    - *Co-locate project parties.* Communication in DB projects is fast and furious, and co-location permits a quicker and more-efficient information flow. Co-located parties should include the DOT project team, consultants, designer, contractors, and any other necessary team member. In the case of the SH 130 project, even the FHWA officer in charge of the project and a representative of the Attorney General's office (to assist in contracts and ROW) were co-located.
    - *Early in the process, allocate responsibilities between DOT employees and engineering consultants.* Define all activities that have to be performed by state employees versus activities that can be performed by owner consultants to help develop an activity flowchart and define the organizational structure. This in turn gives insight into needed resources and assists in planning.
    - *Address the use of Project Management Information Systems in allocating responsibilities between the owner project team and Design-Builder.* Incorporating Project Information Management Systems (PIMS) facilitates project tracking and reduces misleading information. Streamline the process by making the source of the information (Design-Builder) responsible for managing and operating the system, but make the system open to oversight by owner personnel.
- <u>Contract management</u>
    - *Be more prescriptive in critical project specifications.* More prescriptive project specifications ensure independence of the engineering group within the design-build organization. Use of prescriptive specifications should be balanced and thoroughly reviewed by the DOT. For example, if the DOT desires overpasses with no piers in the median (for safety purposes), then this requirement should be given in the specifications. Trying to negotiate this issue later (at a higher price) is not desirable.
    - *Be accurate in defining repetitive pecuniary responsibilities even though of minor concern.* Unclear assignment of pecuniary responsibilities can create harmful and time-consuming "question-answer" loops between contract parties.
    - *Avoid complexity in contract definitions.* Obscure definitions in contract clauses increase the chances for adversarial relationships. For example, the difference between "known" and "unknown" utility locations should be clear and understandable.

- Right of Way (ROW)
    - *Set corridor boundaries to decrease ROW cost escalation.* ROW acquisition costs typically are paid for by the DOT even if ROW procurement services are included as part of the contract (as in the case of SH 130 and SH 45 SE). Defined corridor boundaries prevent the developer from proposing changes to the route strictly to lower construction costs at the expense of higher ROW costs to the owner.
    - *Outline the desired review process of ROW documents in contract clauses.* Usually, DOT employees follow specific procedures for the ROW process. Specifying the desired procedures will avoid inconsistency by the Design-Builder and will facilitate the review process.
- Utility relocation
    - Adopt a prescriptive definition of "utility" and include accurate documentation on existing utilities. An adversarial atmosphere can be generated if the contract parties begin the project without clear definitions of known versus unknown utilities and without accurate utility strip maps.
    - *State clearly the Design-Builder's responsibilities for maintaining a strong relationship with the owner.* Keeping the day-to-day operations on time and on budget demands careful attention to the relationship between the contract parties. Therefore, communication flows between the Design-Builder and the utility owners should be well-defined and carefully followed.
        - The Design-Builder should coordinate the communication process with the major utility owners, and must organize frequent meetings between the utility owners and the project designers. Once research on new easements is underway, utility owners need to be informed on a regular basis of ongoing ROW design developments.
        - Set design review milestones for items in conflict with Utility Adjustments early enough in the process to make new easement acquisition easier for utility owners.
        - Include a process for requesting new utility easements. Requests for easements for new utilities (those not existing at the time of the contract award) notably increased during the initial phases of the SH 130 project. Outlining a process for these situations will decrease time delays and the chance of undermining relationships.
    - *Keep the Quality Assurance (QA)/Quality Control (QC) process simple.* Complex QA checks on utility agreement documentation may have minimal or marginal benefits when the time delays they create are factored into the equation. Start the review process early in order to achieve an optimal trade-off between time and gains.

Conclusions

Use of DB via the CDA can be viewed as an *opportunity* for TxDOT to identify the use and application of DB, and to gain the required knowledge, skills, and experiences needed to implement DB in other areas. Shifting away from the existing paradigm is best achieved by an analysis of how various entities have moved toward a new model of public infrastructure and highway procurement that supports the use of multiple project delivery methods and contracting approaches, as well adopting the lessons learned from SH 130. TxDOT should anticipate that the use of DB will grow as demonstrated by the Department's procurement efforts with the Trans Texas Corridor. As a result, TxDOT should gain the knowledge needed to develop a comprehensive approach of incorporating project delivery methods and contracting approaches to improve highway acquisition and maximize public resources.

As discussed earlier, the first research study outlined in this paper lead to a series of recommendations to assist TxDOT with gaining the full benefits of DB. The following observations are an assessment of which recommendations (*in italics*) were implemented as part of the investigation and documentation of the lesson-learned during the planning and procurement phase of SH 130.

1. Design-build process guidelines and a delivery process (planning, scope, RFP, selection, management, etc.). The delivery process has been drafted as part of the second research project.
2. Assess the availability of the skills required for the use of DB in the organization. These skills are being characterized as part of the second research project.
3. Train selected members of the organization in the use of the DB project delivery system. This task has not been undertaken formally as of yet and most of the learning that has taken place by TxDOT staff has been through interactions with consultants and/or the DB contractor. A series of seminars are planned as part of the second research project to share knowledge across TxDOT.
4. Optimize communication among the parties involved within TxDOT. Co-location has helped facilitate communication between the parties involved with SH 130. This is being studied as part of the second research project.
5. Optimize the pre-project planning process. Developing a lessons-learned process will provide benefit, and having a two-step selection process that allows industry to review and comment on the proposal helps to address issues such as scope problems and risks.
6. Select pilot DB projects that have a relatively certain scope and contain well-known processes and technologies. Although TxDOT has used DB on extremely large projects, these projects are new roads located in mostly rural areas where planning, ROW acquisition, design, and construction are much more predictable than most of TxDOT's portfolio of current projects.
7. Ensure selection of qualified DB contractors. The two-step selection process prequalifies the contractors.
8. *Develop succinct criteria specifications.* Lessons-learned for this topic are being developed as part of the second research project.
9. *Develop a systematic way to evaluate project results to determine if existing DB procedures and approval processes are adequate, and respond to legislative requirements.* This benchmarking process is being developed as part of the second research project.

The review and assessment of other state DOTs in implementing similar processes shows that success has been achieved by organizations that have been proactive in their approach to managing the transition. The lessons learned from the development of SH 130 are an additional asset to help TxDOT accommodate and effectively undertake innovative procurement and contracting practices. The following conclusions are to help TxDOT in its quest to identify the factors that can inhibit efforts to improve project quality, cost, and schedule.

- TxDOT employees and others involved with innovative project delivery methods and contracting approaches need adequate training in the DB contracting process to understand and perform the duties required of public owners.
- TxDOT and others should develop a systematic method for capturing project performance data that can be used to monitor the impacts on implemented changes and respond to legislative reporting requirements.
- Implementing innovative project delivery methods and contracting approaches is a paradigm shift that requires a commitment from staff and senior management to accept the challenge and provide adequate leadership.

References

American Association of State Highway and Transportation Officials. 1998. Primer on Contracting 2000 (1998). http://www.aashto.org/info/primer/primer_1-18.html. Accessed January 2001.

Associated General Contractors (AGC) of America-Texas Building Branch, Texas Society of Architects, and Consulting Engineers Council of Texas, Inc. 1998. Project Delivery for Texas Public Schools. A publication of the AGC-Texas Building Branch, Texas Society of Architects, and Consulting Engineers Council of Texas, Inc., Austin, TX.

Federal Highway Administration. 1998. Initiatives to Encourage Quality Through Innovative Contracting Practices Special Experimental Projects No. 14 (SEP-14). U.S. Department of Transportation, Federal Highway Administration.

Federal Highway Administration. 2000. SEP-14 Design-Build Information Web Site. http://www.fhwa.dot.gov/infrastructure/progadmin/contracts/d_build.htm FHWA Web Site. Accessed December 2000, and May 2001.

Federal Highway Administration. 2001. Status of FHWA Design-Build Rule Making, FHWA Briefing Paper, Dated May 18, and provided by the FHWA Infrastructure Business Unit.

Federal Highway Administration (FHWA) (2002). "Design-Build Contracting; Final Rule". Federal Register, 67(237), pp. 75901-75935.

Gallegos, G. 2001. "University of Texas System Design-Build Practices," Presentation at the Design-Build Institute of America (DBIA)-Texas Chapter Meeting, January 31, 2001, Austin, Texas.

Gibson, G. E. and Walewski, J. 2001. "Project Delivery Methods and Contracting Approaches: Assessment and Design-Build Implementation Guidance," Research Report Number 2129-P1, Center for Transportation Research, August.

Gibson, G. E. and Walewski, J. 2002. "Project Delivery Methods and Contracting Approaches Available for Implementation by the Texas Department of Transportation," Project Summary Report Number 2129-S, Center for Transportation Research, April.

Hancher, D. 1999. "Innovative Contracting Practices," TR News, Number 205, Nov-Dec 1999.

Loulakis, M. C., and Huffman, R. D. 2000. "Project Delivery and Procurement: Understanding the Differences," Presentation Summary from the DBIA/AIA Professional Design-Build Conference, October 5, San Diego CA.

Migliaccio, G. C., Gibson, G. E. and O'Connor J. T. 2005. "Delivering Highway Projects Through Design-Build: Analysis of the Comprehensive Development Agreement (CDA) Procurement Process in Texas," *Proceedings of the Construction Research Congress 2005*, ASCE, San Diego, CA, April 2005.

National Cooperative Highway Research Program. 1991. Innovative Contracting Practices. Transportation Research Board Task Force on Innovative Contracting Practices (A2T51). TRB Document Number C386.

National Cooperative Highway Research Program. 1996. Constructability Review Process for Transportation Facilities. Project 10-42. Texas Transportation Institute.

National Cooperative Highway Research Program. 1999. Draft Guidebook to Highway Contracting for Innovation: The Role of Procurement and Contracting Approaches in Facilitating the Implementation of Research Findings. Transportation Research Board, National Research Council, Washington, D.C.

O'Connor J.T., Gibson G.E. Jr., Migliaccio G.C. (2004a). "Product No.1 – CDA Procurement Process Model", Research Report Number 0-4661-P1, Center for Transportation Research, The University of Texas at Austin, Austin, Texas.

O'Connor J.T., Gibson G.E. Jr., Migliaccio G.C. (2004b). "Product No.2 - Essential Elements of CDA Master Contract", Research Report Number 0-4661-P2, Center for Transportation Research, The University of Texas at Austin, Austin, Texas.

O'Connor J.T., Gibson G.E. Jr., Migliaccio G.C. (2004c). "Research Project 0-4661, R1- 2004 Annual Interim Report", Research Report Number 0-4661-1, Center for Transportation Research, The University of Texas at Austin, Austin, Texas.

Pensock, Edward. 2004. Comprehensive Development Agreements & Public Private Partnerships in Texas. Presentations at the Partnerships for Transportation Workshop held in Sacramento, California on October 20, 2004.

Schenck, J. S. 2000. Design-Build: Introduction. Design/Build: A Guide to Licensing and Procurement Requirements in the 50 States and Canada. John R. Heisse, Editor. American Bar Association.

Schrank, D., and Lomax, T. 2004. Texas Transportation Institute, "2004 Annual Urban Mobility Report".

Texas Comptroller. 2000. eTexas: Recommendations from the Texas Comptroller. http://www.e-texas.org/recommend/. Accessed January 2001.

Texas Department of Transportation (TxDOT). 2000. Contract Administration Handbook for Construction Projects, "Daily Road-User Costs, Incentives and "A+B" Bidding."

TxDOT. 2001. Request for proposals to construct, maintain and repair the SH130 turnpike through an exclusive development agreement: Exclusive Development Agreement. Austin, TX: Texas Department of Transportation.

TxDOT. 2004. Request for detailed proposals to develop, design, and construct the SH45 Southeast turnpike through a comprehensive development agreement: Comprehensive Development Agreement. Austin, TX: Texas Department of Transportation.

TxDOT. 2005. Central Texas Turnpike System 2002 Project: Monthly Activity Report, Period ending March 31, 2005. Austin, TX: Texas Department of Transportation. http://www.texastollways.com/tta/pdfs/April%20Final%20View.pdf (accessed on 07/07/2005)

Vernon's Texas Statutes and Codes Annotated. 1999. St. Paul, Minn.: West Publishing Company.

Walewski, J., Gibson, G. E. and Jasper, J. 2001. "Project Delivery Methods and Contracting Approaches Available for Implementation by the Texas Department of Transportation," Research Report Number 2129-1, Center for Transportation Research, October.

Appendix A
Matrix of Project Delivery Methods
(Source: Walewski, Gibson, and Jasper 2001)

Table 2.3 Matrix of Project Delivery Methods

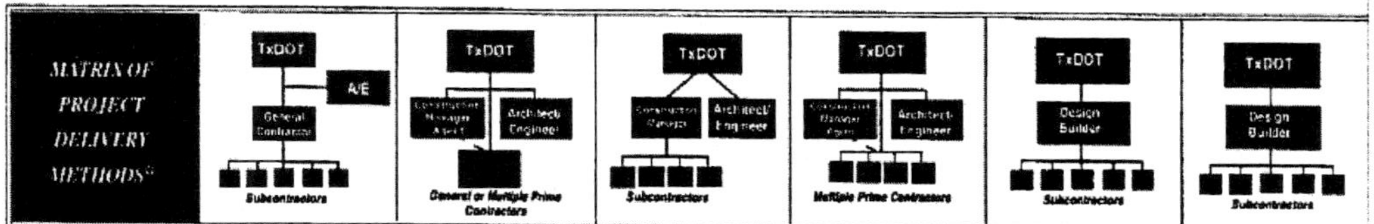

Delivery Method	Traditional Process, Design-Bid-Build	Construction Manager-Agent	Construction Manager-at-Risk	Multi-Prime	Design-Build	Design/Build/Operate
Definition	A delivery method where TxDOT selects an architect/engineer based on qualifications to design and develop construction documents from which TxDOT solicits lump sum bids. Selection of the contractor is based on the lowest responsible bid and the contractor serves as a single point of responsibility for construction.	A method where the construction manager serves as an agent for the owner providing pre-construction and construction services in lieu of a general contractor. The construction manager-agent provides design phase assistance but holds no subcontracts nor provides project bonding for the construction. A GC or multiple trade contracts are held by the owner. Selection is based on the proposal offering the best value to the owner.	A method where the construction manager serves as the general contractor providing pre-construction and construction services. The construction manager-at-risk provides design phase consultation in evaluating costs, schedule and implications of alternative designs. A guaranteed maximum price (GMP) may be issued and the CM-at-risk serves as the responsible party contracting directly with subcontractors during construction	A method where the owner, or sometimes an agent, oversees multiple contractors as opposed to a general contractor conducting total oversight. The owner or agent assumes greater control over the project but also assumes significantly more risk. The multiple trade contracts are usually held by the owner. Selection is based on the proposal offering the best value to the owner.	A method where a single entity is contracted to provide both design and construction. The design-build team consists of contractor and architect/engineer who contract directly with the subcontractors and is responsible for delivery of the project. Selection of the design-build contractor is based on the proposal offering the best value.	A form of design/build where the investment of private capital is used to finance, design, construct, operate, and maintain a road or highway project for public use for a specific term. During the term the investor are paid-back with toll revenue and after an agreed upon time the project reverts to the public owner.
Pros	• Familiar delivery method • Defined project scope • Single point of responsibility for construction • Open, aggressive bidding • Limits graft and corruption	• Design phase assistance • Selection flexibility • Faster schedule delivery • Change flexibility • Non-adversarial relationship	• Selection flexibility • Design phase assistance • Single point of responsibility for construction • Team concept • Faster schedule delivery • Change flexibility	• Selection flexibility • Cost savings possible • Faster schedule delivery • Change flexibility • Greater control over project	• Selection flexibility • Single point of responsibility for design and construction • Faster schedule delivery • Team concept	• Single point of responsibility for all project components • No up-front public cost • Risk carried by investors • Faster schedule delivery • Life-cycle design
Cons	• No design phase assistance • Longer schedule duration • Price not established until bids • Adversarial relationship • Lack of flexibility for change	• No single point of responsibility • No guaranteed price • Owner must manage many contracts	• Adversarial relationship reduced • Difficult for owner to evaluate GMP	• No single point of responsibility • No guaranteed price • Owner must manage many contracts	• Loss of check and balance • Different management techniques required • Potential adversarial relationship between owner and Design/ Builder	• Loss of check and balance • Investment decisions rule • Difficult process to manage • Limited to toll roads
Best Suited	New projects that are not schedule sensitive or subject to potential change.	Large new or renovation projects that are schedule sensitive, difficult to define, or subject to change.	Larger new or renovation projects that are schedule sensitive, difficult to define, or subject to change.	Larger new or renovation projects that are schedule sensitive, difficult to define, or subject to change.	New or renovation projects that are schedule sensitive	Larger new projects that lack adequate public funding.
Least Suited	Complex projects that are sequence or schedule sensitive. Projects subject to potential change.	Smaller projects	Smaller projects	Smaller projects	Projects that are difficult to define, and are less schedule sensitive.	Smaller projects and those that are not investment grade.

Development of Performance Warranties for Performance-Based Road Maintenance Contracts

Mehmet Egemen Ozbek[1] and Jesús M. de la Garza[2 3]

Abstract

Performance-based contracting in the transportation arena is a rather new concept which has a few applications in the United States. The Virginia Department of Transportation (VDOT) has been one of the first state agencies to take the initiative of using a performance-based contract for the maintenance of a portion of its interstate highway system. This initiative led to the establishment of a contract between VDOT and a private contractor in 1996.

Just like performance-based contracting, the use of warranties in highway construction contracts in the United States is quite a new concept, which mainly dates back to 1995. There are different views possessed by different parties about the possible outcomes of implementing warranties in highway contracts.

This research first explores the benefits that would/may be gained by the incorporation of warranty clauses into performance-based highway maintenance contracts, discusses the rationale behind developing warranty clauses for such contracts, and then develops a warranty clause template and proposes that it be used in the future performance-based road maintenance contracts issued by VDOT.

Introduction

In July 1995, the Public-Private Transportation Act (PPTA) of Virginia was passed. This act authorized the Commonwealth of Virginia (responsible public entity) to establish contracts with private entities to acquire, construct, improve, maintain, and/or operate one or more transportation facilities within the state of Virginia. PPTA further states that this approach of privatization may result in the availability of such transportation facilities to the public in a more timely or less costly fashion (Code of Virginia §56-558 A-3). Three months after PPTA was passed, in October 1995, a contractor submitted an unsolicited proposal to VDOT for the maintenance of assets within a portion of the interstate highway system of Virginia. After an intensive evaluation of this proposal by VDOT (including the phases of a detailed proposal re-submittal by the contractor and negotiations), in December 1996, the "*Comprehensive Agreement for Interstate Highway Asset Management Services*" was signed. The 5-year contract required the contractor to administer and maintain all assets, and carry

[1] Graduate Student, Charles E. Via, Jr. Department of Civil and Environmental Engineering, Virginia Tech, Blacksburg, Virginia.

[2] Vecellio Professor, Charles E. Via, Jr. Department of Civil and Environmental Engineering, Virginia Tech, Blacksburg, Virginia.

[3]Program Director, Information Technology and Infrastructure Systems, Civil and Mechanical Systems Division, National Science Foundation, Arlington, Virginia.

out incident management and snow removal facilities, on a total of 250 miles (nearly 25% of Virginia's interstate highways). The contract was renewed in June 2001 for 5 more years (contract term to commence in July 2002 and end in July 2007) with contract provisions (other than the fee) remaining exactly the same as the previous terms.

A very important aspect of the contract is its performance-based nature. A performance-based contract is very different from a method and/or material properties based contract (traditional contract types which are common in the construction industry) in one great sense. A performance-based contract, as the name implies, sets forth the performance expected from the end product of a project, rather than directing the contractor with the methods to achieve that end product. In other words, a performance-based contract specifies the desired outcomes rather than the desired processes to reach those outcomes. According to Frost and Lithgow, a performance-based road maintenance contract is "a contract under which the contractor undertakes to plan, program, design, and implement maintenance activities in order to achieve specified short and long term road condition standards for a fixed price, subject to specified risk allocation" (Frost and Lithgow 1998). In addition to being a performance-based contract, the current VDOT contract is an asset management contract. In fact, the contract is for the "Interstate Highway Asset Management Services" as is indicated on it. As Falls, Haas, McNeil, and Tighe assert, asset management is "a systematic process of maintaining, upgrading, and operating physical assets cost-effectively. In the broadest sense, the assets of a transportation agency include physical infrastructure such as pavements, bridges..." (Falls et al. 2001).

This research deals with performance-based contracts that cover the maintenance of assets within the highway right-of-way, i.e. from fence to fence. The research first explores the benefits that would/may be gained by the incorporation of warranty clauses into performance-based highway maintenance contracts, discusses the rationale behind developing warranty clauses for such contracts, and then develops a warranty clause template and proposes that it be used in the future performance-based road maintenance contracts issued by VDOT.

Definition of Terms

This section describes the three main terms (performance criteria, performance targets, and an asset item's overall condition rating) that are related to the annual inspections performed as a part of the abovementioned contract. It is essential for the reader to understand these terms as they are used throughout the course of the discussions within this paper.

The performance criteria, as used within the contract, are established for each asset item. They are the measures established to assess the condition of the assets. There may be more than one performance criterion established for each asset item. For example, for the "Paved Ditches" asset item, one of the established performance criteria is to have a maximum of 1-inch settlement. Another one is having a maximum of 25% of the entire area of the paved ditch within a sample unit as spalled.

The performance targets, as used within the contract, are established for each asset item. For a certain asset item, its performance target is the percentage that defines the ratio of the minimum amount of the asset item that is required to meet the performance criteria, to the total amount of that asset item. For example, the performance target for the "Paved Ditches" asset item is 90%.

For any asset item, the asset item's overall condition rating is the ratio (in percentage) of the number of the asset item (among the total inspected amount) meeting the performance criteria, to the total number of that asset item inspected in the annual inspections. For example, within an inspection, if out of the 100 inspected paved ditches, 80 are assessed as meeting the performance criteria, the "Paved Ditches" asset item's overall condition rating is 80% for that inspection. However, the findings of these inspections can be generalized to the whole population of that asset item by following a robust statistical process through randomly choosing the number of samples to be inspected for the asset item. This phenomenon enables VDOT and the independent consultants to avoid inspecting all of the assets. Hence, for the purposes of this research, the term "asset item's overall condition rating" means "the ratio (in percentage) of the number of the asset item meeting the performance criteria, to the total number of that asset item".

Problem Definition

Some of the greatest benefits that performance-based contracting can bring to a project are the innovations implemented into the materials and processes that are used by the contractor. In some cases, innovations may also bring some undesirable consequences to the projects. Using performance specifications gives the contractor an increased control over the project. Hence, there should be a kind of assurance so that the increased control of the contractor and innovative techniques implemented by him will yield to a service and/or product which is sought by the agency and which also has a predefined quality. In other words, there should be an assurance to balance the enhanced opportunity for the contractor input and control by a greater assumption of liability by the contractor (MDOT 2002).

One of the ways of providing this assurance is through the implementation of warranties. Warranties may provide the owner with some kind of an assurance for the quality of the services and/or products to be received. The warranty concept (which will be discussed later on in detail), if implemented carefully, can ensure that performance-based contracting would lead to quality services and/or products which have long term performances. In plain language, a warranty may act as a buffer for agencies to overcome performance-based contracting's aforementioned vulnerability to the issues of innovation and control.

The existing performance-based contract calls for once per year inspections (performed at a prefixed schedule) to assess the contractor's performance in meeting the predetermined performance criteria. During these inspections, services and products obtained as the results of contractor's maintenance work are evaluated. This is done by checking the conditions of all asset items against certain VDOT-established performance criteria. Then, overall condition ratings for each asset item are compared to VDOT's established performance targets. In such a scheme, the contractor is said to meet the contract requirements for an asset item when the results of the inspection reveal an overall condition rating (for the asset item) which exceeds the VDOT established performance target for that asset item.

A hypothetical example with an illustration is presented in Figure 1. This figure shows the percentage of the "Paved Ditches" asset item that meets the performance criteria during the contract period. As can be seen in Figure 1, for the "Paved Ditches" asset item, the established target is: "Meeting the predetermined performance criteria for 90% of the paved ditches". There is a deterioration curve for the "Paved Ditches" asset item. And, as it deteriorates and its overall condition rating falls below the performance target, it is maintained by the contractor to be pulled to the performance target. As can be observed in Figure 1, the contractor is trying to meet the performance criteria for the amount of paved ditches that makes up the 90% (performance target) of all paved ditches.

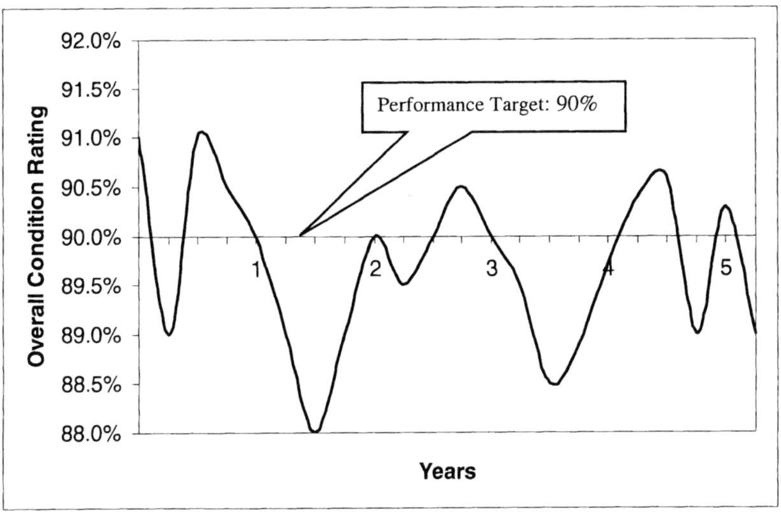

Figure 1. Contract without warranty

This is the kind of maintenance philosophy that can easily be implemented by a prospective contractor doing work under the current contract clauses. This maintenance philosophy provides the desired level of performance as per the contract, but it does not improve the quality and overall condition of the asset items. Moreover, it does not ensure the long-term performance of the assets that are maintained. This significantly contradicts the concept of asset management that requires that the assets maintained by the contractor be upgraded (Falls et al. 2001). This also contradicts the definition of performance-based road maintenance contracts that calls for the achievement of long term road condition standards (Frost and Lithgow 1998).

It is essential to comprehend what is meant by the concepts of "improvement of the overall condition of assets" and "long term performance of assets". The assets within the highway system are continuously in use and thus deteriorating. There is a point in their life cycle at which no more deterioration is acceptable due to the safety and/or aesthetic reasons. This is defined as the threshold level by the agencies. Long-term performance means that the time for these assets to reach that particular point (threshold level) should be long enough as dictated by the agency. And, this longevity can be attained by improving the asset's condition. During the maintenance activities, if the overall condition rating of the asset item is barely kept at the level stated in the performance targets, then as the maintenance activities are terminated, that asset item's overall condition rating drops below the acceptable performance target level in a very short time. This discussion is represented in Figure 2 graphically, again with hypothetical values.

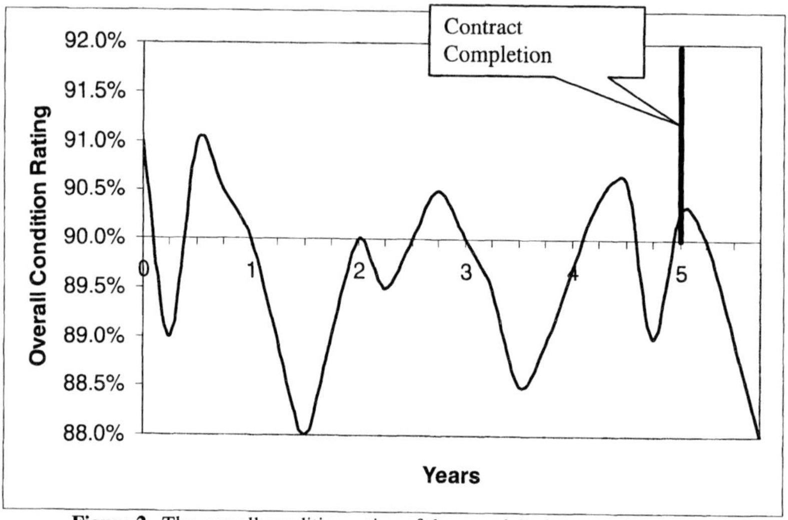

Figure 2. The overall condition rating of the paved ditches asset item before and after the contract completion date

Benefits of and the Rationale behind Developing Warranty Clauses

The existing contract conditions and clauses neither force nor encourage the contractor to implement a maintenance philosophy which would result in the improvement in the overall conditions of the assets and which would provide the long term performance of the assets. Under the current contract language, the contractor is only obligated to achieve the performance criteria and performance targets without improving their overall condition or providing long term performance.

As mentioned earlier, the usage of warranties in highway contracting is a rather new concept and thus there is a very limited amount of data gathered about the implementation of warranties to date. Different state DOTs possess different views with respect to different issues (i.e. cost, bonding, litigation etc.) as far as the advantages and disadvantages of warranties are concerned. However, it is reasonable to assume that warranty provisions covering the issue of defects could potentially serve as built-in incentives for the contractor to strive to ensure the long term performance of the assets. This is mainly due to the fact that under such a provision, the contractor would otherwise incur the costs at a later time because of the warranty it provides.

Warranty clauses for performance-based road maintenance contracts, if developed carefully, may give the contractor the motivation to implement superior maintenance practices to improve the overall conditions of the asset items during the contract period, as illustrated in Figure 3 using hypothetical values. This is done by lengthening the life of each asset item and by achieving long term performance; otherwise the contractor is required to come back and implement remedial actions to the asset items showing defective properties to bring them back to or above the warranty criteria required throughout the warranty period. By contrasting Figure 2 and Figure 3, one can see that the performance of an asset item as depicted in Figure 3 is more desirable than the one depicted in Figure 2.

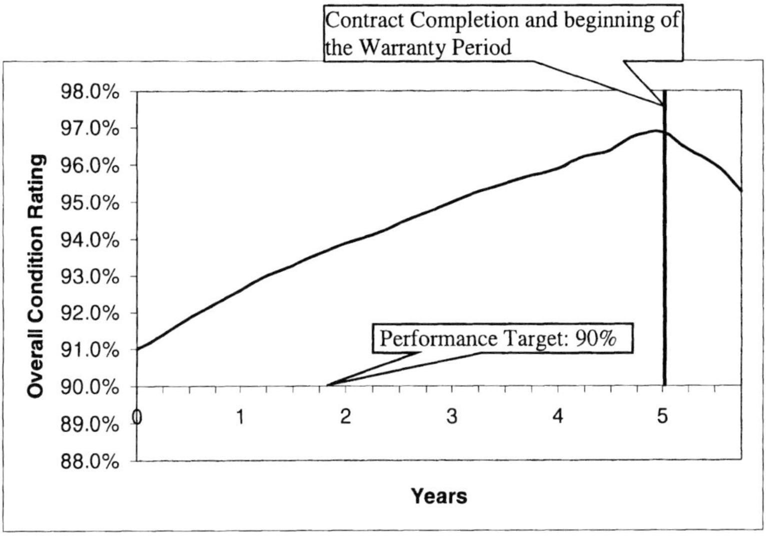

Figure 3. Contract with warranty

Methodology and Literature Review

According to Burrell and Morgan, research methodologies can vary in a range. However, they define the research methodologies at the extreme ends of the continuum as nomothetic and ideographic. Ideographic methodology emphasizes the analysis of subjective ideas which the researcher proposes. In this methodology, the researcher, observing the situations and involving her/himself in the everyday flow of life, generates these ideas (Burrell and Morgan 1979). The methodology used in this research is an ideographic one. Therefore, this research possesses the characteristics of a qualitative research as opposed to a quantitative one, in which a very scientific strictness and standard is present in the process of testing the validity of the hypothesis of the research.

Due to the fact that this research has a qualitative nature, the researchers, after observing and analyzing the situations related to the subject matter through an intense literature review, proposed subjective ideas they believe to be the best solution to the identified problem. However, as an essential part of the qualitative research, these ideas needed to be evaluated by some experts in order to minimize (or at best to completely avoid) the subjectivity, which may act as a major drawback present in the achieved solution. Due to the content of this research, these experts needed to be individuals who are experienced and knowledgeable in the warranty field. Therefore, after the warranty language was developed, it was presented to a panel of experts, members of which were professionals who are highly involved in highway construction and maintenance projects, and who also have sizeable knowledge in the administration of warranties. The panel basically was composed of professionals representing parties involved in highway construction and/or maintenance projects (i.e. the contractor and the state DOT,) and professors within academia that have been performing research about the warranty concept for a long time.

In performing this research, a fair amount of time was spent in identifying and going through the literature, which was examined under the following subject headings:

- Introduction and development of highway warranties in the USA
- Examination of highway warranties in the USA
- Highway warranties in European countries
- Examples of performance-based road maintenance contracts with warranties
- Performance-based road maintenance contracts issued in the USA
- Standard forms of traditional design-bid-build contracts
- Highway and bridge specifications prepared by the state DOTs
- Examples of performance-based contracts other than road maintenance contracts issued in the USA

Within this literature review, special consideration was given to the examination of highway *construction* warranties in the USA. In fact, while developing the warranty clause template, the boilerplate languages that are present in a number of state departments of transportation (DOTs) construction warranty clauses have been used to some extent. The states whose warranty clause languages were used as boilerplates are Wisconsin, Ohio, Minnesota, and Florida. Within this group, Wisconsin and Ohio were particularly chosen due to their long- time experiences with the warranty concept. Moreover, the model asphalt warranty specification that has been developed by Anderson and Russell (Anderson and Russell 2001) as a part of a NCHRP Project (NCHRP Project 10-49) was also used.

It is important to note that all of the abovementioned warranty clauses are for the "construction" of highway elements as opposed to the "maintenance" of such elements. Furthermore, most of the abovementioned warranties are material and workmanship warranties (for traditional design-bid-build contracts) as opposed to the performance warranty (for performance-based contracts), which is the type of warranty developed in this research. Hence, rather than solely depending on the abovementioned clauses, the researchers developed the warranty clause template by means of adopting the principles and the language that were used in the abovementioned warranty clauses, and then adapting those to meet the specific needs of the contract type that is in the scope of this research, i.e. the performance-based road maintenance contract. Moreover, some new concepts and approaches were developed to establish the most important components (the warranty criterion and the length of the warranty period) of the warranty clause template, which will be discussed in the following sections.

Other than the abovementioned states whose warranty clause languages were used as boilerplates, many other state DOTs' experiences with the warranty implementation as well as the Federal Highway Administration's perspective on the warranty concept were investigated as sources of further information. The warranty experiences of the following states were investigated: Michigan, Kansas, Missouri, New Mexico, South Carolina, California, Indiana, Montana, Colorado, and North Carolina.

Approach Developed for Defining Defects and Establishing the Warranty Criterion for Each Asset Item

For a given asset item, defining defects is essential to develop the warranty criterion for that asset item. This is mainly due to the fact that what VDOT wants to include in the warranty criterion (which will be used in assessing the warranty compliance) as an indicator of poor performance or poor overall condition of an asset item is totally dependent on what it perceives as a defect resulting from the unacceptable work of the contractor.

Defects for each asset item can be characterized in terms of the performance criteria and performance target established for that asset item. For example, by examining Table 1 for the "Paved Ditches" asset item, the defect can be defined as "*meeting the performance criteria* for **less than 90% of the total amount of paved ditches present in the interstate system maintained by the contractor**". It can be noted that the first part of this statement (presented in *italics)* is established using the performance criteria and the second part (presented in **bold**) is established using the performance target.

Thus, by applying the same logic, the broadest definition of the defect for the "Paved Ditches" asset item can be stated as "(a) having less than 1 inch of settlement or (b) having no undermining or undercut that requires action or (c) having less than 25% of the area spalled or (d) having no obstruction to flow of water that requires action or (e) any combination of (a), (b), (c) and (d); for less than 90% of the total amount of paved ditches present in the interstate system maintained by the contractor".

Table 1: Performance Criteria and Performance Target for the "Paved Ditches" asset item

Asset	Performance Criteria	Performance Target
Paved Ditches	• < 1'' settlement • no undermining or undercut requiring action • < 25% spalled • no obstruction to flow of water that requires action	90%

It is important to note that the discussion presented above is applicable for the duration of the contract term. In other words, in preparing the contract, VDOT established the performance criteria and performance targets (both of which are used in the definition of the defect for each asset item) for the contract term. However, if VDOT perceives "meeting certain performance criteria for less than a pre-defined target for an asset item" as a defect for the contract term, the same definition of that defect can be applicable after the contract term too. To state it more clearly, the performance criteria and targets are established with the long time experience of VDOT, and hence, an asset item perceived to be defective during the contract term when maintenance work is being performed, may well be perceived to be defective after the completion of the contract term, if it possesses the criteria and overall condition rating that make it defective. Therefore, there is no point in generating two different definitions of the defect: One for the contract term, and one for the warranty period.

By utilizing the definition of the defect as established above, the warranty criterion for the "Paved Ditches" asset item can be established as "*meeting the performance criteria* for **at least equal or more than 90% of the total amount of paved ditches present in the interstate system maintained by the contractor**". Using the previous example, the warranty criterion (in its broadest sense) for the "Paved Ditches" asset item can be stated as "(a) having less than 1 inch of settlement and (b) having no undermining or undercut that requires action and (c) having less than 25% of the area spalled and (d) having no obstruction to flow of water that requires action; for at least equal or more than 90% of the total amount of paved ditches present in the interstate system maintained by the contractor".

Approaches Developed for Establishing the Length of the Warranty Period

Alternative 1: In this alternative, the normal deterioration curve of the asset item is used to identify the time necessary for that asset item to deteriorate from its new state to the state where it is said to be defective. Such a curve should reflect the situation in which the asset item deteriorates from normal daily usage without being maintained. Moreover, this curve should be plotted for only normal daily conditions in which there is not any extreme event such as very severe weather and/or very heavy traffic. Figure 4 illustrates a hypothetical normal deterioration curve for the "Paved Ditches" asset item. The term "new state" means the state in which the "Paved Ditches" asset item is in its best possible condition, i.e. newly constructed. This implies that the overall condition rating for the "Paved Ditches" asset item is 100%, meaning that all of the paved ditches present in the interstate system are meeting the performance criteria. The "defective state" for the "Paved Ditches" asset item presented in the graph is exactly the state at which the "Paved Ditches" asset item was defined to be defective, i.e. the pre-established performance criteria are met for less than the performance target.

As discussed above, this curve reflects the situation in which no maintenance activity is performed. It would be fair to assert that an asset item, which is well-maintained during the contract term, should reach its best possible condition (i.e. its new state) by the completion of the contract term. And it would also be fair to state that upon the completion of the contract, and thus with the beginning of the warranty period, the same well-maintained asset item would begin to deteriorate in accordance with its normal deterioration curve and thus would reach its defective stage at the time that is specified in its normal deterioration curve. And given all of these conditions, it would be fair to assert that the time in which the well-maintained asset reaches its defective stage marks the end of the warranty period. Note that this time is represented as "X" in Figure 4. In conclusion, the warranty period may be established as "the period starting with the completion of the contract term and ending with the time which reflects the defective state of the asset item in the normal deterioration curve of that asset item".

Alternative 2: The above- proposed alternative of establishing the warranty duration has a major drawback when examined thoroughly within the highway maintenance concept. As presented above, Alternative 1 assumes and requires that at the end of the warranty period, as the maintenance activities are terminated, the asset item's overall condition rating should be 100%. This implies that the maintenance activities performed during the contract term should be in such a way that they should improve the asset item's overall condition to the maximum extent possible. This means that the asset item should be brought up to its new state, making it as good as newly constructed. However, expecting the contractor to improve the asset item's overall condition to the maximum extent possible to reach an overall condition rating of 100% is both unreasonable and unrealistic given the nature of the highway maintenance projects. To resolve this issue, the length of the warranty period, which was established previously in Alternative 1, should be reconsidered. The warranty period can be re-established using a different portion of the normal deterioration curve. Alternative 2 determines the length of the warranty period by using a normal deterioration curve in which the initial overall condition rating of the asset is assumed to be at a certain level above the performance target but below 100% (depicted as "A" in Figure 4). In this case, the time for the asset item to deteriorate from its state represented by "A" to the defective state is represented by the parameter "Z".

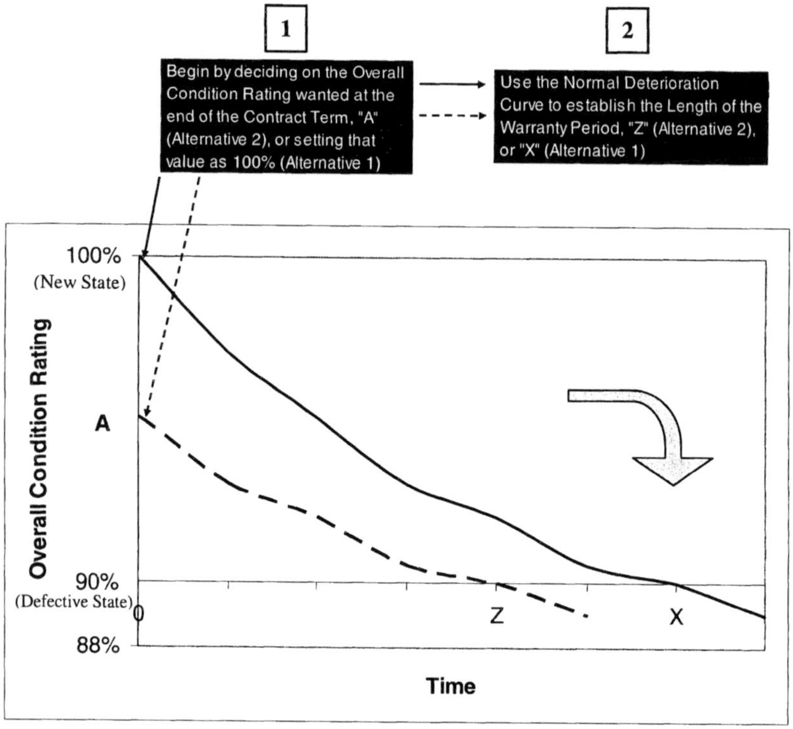

Figure 4. Establishing the length of the warranty period using Alternative 1 or Alternative 2

Alternative 3: Alternative 3 calls for the establishment of the warranty period without defining any intermediate value like "A" that helps to calculate the length of the warranty period. This alternative proposes that the length of the warranty period should be determined by the state agency, VDOT, according to its expectations with regards to the performance of the highway and its assets. For this, VDOT needs to set the time period in which it would not be willing to maintain the asset item or to pay for the maintenance of the asset item after the completion of the contract term. This time period ("W"), in fact, constitutes the warranty period it wishes to implement. Then, to ensure that this established length is reasonable, back calculation should be made (using the normal deterioration curve as illustrated hypothetically in Figure 5) to identify the value of the overall condition rating required at the end of the contract term ("B") for the contractor to be able to meet the warranty criterion all through this established warranty period.

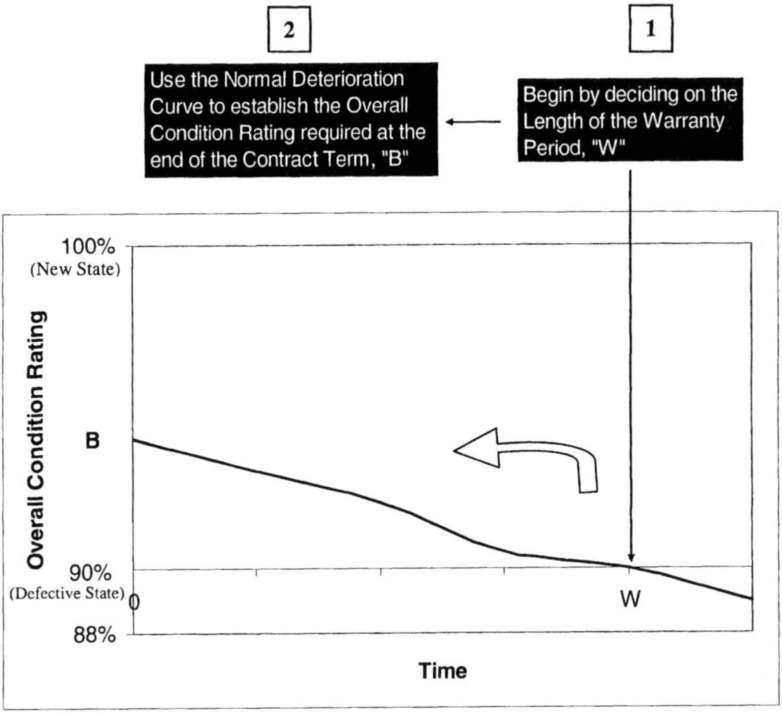

Figure 5. Establishing the length of the warranty period using Alternative 3

After the abovementioned alternatives were developed, they were presented to the panel of experts (consisting of individuals from academia, the construction industry and VDOT) to receive feedback on the practicability of using them, as well as the advantages of each one on the others. The majority stated that Alternative 1 is unrealistic given the requirement of bringing an asset item to its new state at the end of the contract term. The panel of experts also agreed on the impracticality of Alternative 2 due to its vast dependence on the normal deterioration curves and unavailability of good data to establish such deterioration curves. The panel of experts has deemed Alternative 3 as the most viable one. The researchers have presented the alternatives and the feedback of the panel of experts to VDOT. If VDOT chooses to implement the developed warranty clause in its future contracts, it has to decide on which alternative to proceed with, based on its preferences. If VDOT is more interested in the overall condition improvement that should be performed during the contract term and wishes to define the asset item's overall condition rating that it is willing to get at the end of the contract term, it should choose Alternative 2. On the other hand, if VDOT is more interested in setting the time period in which it would not be willing to maintain the asset item or to pay for the maintenance of the asset item after the completion of the contract term, it should proceed with Alternative 3.

The Developed Warranty Clause Template

It is important to note that the researchers propose that different warranty clauses need to be developed for each asset item within the highway right-of-way. This is mainly due to the fact that each asset item is unique and has its own properties defining its lifetime. Each asset item is exposed to different conditions causing it to deteriorate to different extents when compared to other asset items. Moreover, some asset items have relatively higher importance than others as far as the road users' safety is concerned. For the purposes of this research, only a warranty clause template is developed. The researchers believe that once the language for the warranty clause template is developed, the warranty clauses for each asset item can be developed by studying the components of this warranty clause template. The warranty clauses developed for each asset item need to have a common and standard format, and thus a common and standard language. However, in order to compensate for different natures of asset items, the content of each developed warranty clause needs to reflect the properties of the asset item for which it is developed.

The developed warranty clause template is designed to act as a part of the performance-based road maintenance contracts issued by VDOT as opposed to acting as a separate legal instrument as stressed by the attorneys within the state of Michigan (Galehouse 1998). Thus, as warranty clauses are developed for each asset item, they should be incorporated into the contract as new documents, perhaps under the main section, i.e. Article XVI, which has the heading, "Warranties" (current contract has 15 main sections). Given this fact, it can be asserted that all the statements that are made within the earlier 15 sections (Articles) as well as in the following Exhibits of the contract are valid for the warranty period too. In other words, all the rights and the responsibilities that VDOT and the contractor are said to possess as set forth in the previous 15 sections (Articles) as well as in the following Exhibits are to remain in effect during the warranty period too. This would assure that the developed warranty clauses are incorporated into the contract within a whole new Article, as to be a part of the contract and to possess the general tone of the Articles coming before it, as well as the Exhibits coming after it. Thus the warranty clauses will neither change nor affect, and thus interfere with, any other provisions and clauses that are already present in the contract.

The warranty clause template, which is presented in the Appendix, is developed without considering the asset item for which it is developed. It is merely a template which possesses all of the components of a warranty clause. As can be observed, this clause contains some blanks that need to be filled in by considering the properties of the asset item it is developed for. Hence, the language and the format of this template can be used to develop the warranty clauses for each asset item.

Issues at Large

The warranty concept is not a simple solution to the issues posed by performance-based highway maintenance contracting as presented in this research. If a state DOT is to implement warranties for its performance-based highway maintenance contracts, it needs to address many issues that come along with the warranty concept. In other words, the warranty concept is not a panacea that would enable state DOTs to acquire the best value from their performance-based maintenance contracts, just by its implementation. Some of the issues that need to be elaborated on before proceeding with the warranty concept can be listed as: the type of projects appropriate for warranty implementation, the length of the warranty period, the criteria to be used in evaluating warranty compliance, the evaluation of proposals and award of the contract, evaluating warranty compliance and implementation of remedies, bonding issues, cost issues, quality issues, disputes and litigation, quality assurance/quality control, and data collection and inspection by the state DOT.

Even for the same issues like cost, quality, and many others, different sources cite different perceptions of the warranty concept (as being advantageous or disadvantageous) by different parties. For example, the results of the survey conducted by Adam Ross with the attendance of 5 state DOTs reveal the diverse perceptions about the warranty concept (Hancher 1999). In this survey, the benefits received through the implementation of warranties were rated as 3.2 in a scale in which 1.0 represented low and 5.0 represented high. Similarly, difficulty of implementation was rated as 3.0 in the same scale (Hancher 1999). The ratings given by the state DOTs to both of the criteria indicate the DOTs have not reached a specific conclusion about the warranty concept yet. A reason for this may be the fact that the warranty is a rather new concept. Thus, entities have insufficient and different experience with it to date. Therefore, definite conclusions, as agreed upon by everyone, cannot be drawn yet. This will be possible through the implementation of warranties in a wider range of highway projects by a wider range of state DOTs. Another reason for having counter ideas for exactly the same issues may be the different approaches of entities to the warranty concept in different states. The states, which are aware of the possible drawbacks of the warranties, may be cautious of such issues and then convert these drawbacks to benefits. Other states, however, may not be doing this and thus may perceive that the warranties possess more drawbacks than advantages.

Given these conditions, it is a worthwhile effort for the state DOTs to implement warranties in performance-based maintenance projects, at least for a trial basis on pilot projects, in order to gain a better understanding. By this, the state DOT can experience the advantages and disadvantages that the warranties bring about and may then decide whether to proceed with warranties or not.

Conclusions

This research developed a warranty clause template that is proposed to be used in future performance-based road maintenance contracts issued by VDOT. The main rationale behind developing this warranty clause template is to make the prospective contractors implement a maintenance philosophy that would provide improvement in the overall condition of each of the asset items they maintain during the contract term, rather than just maintaining them to meet the performance targets specified in the contract. More importantly, this maintenance philosophy is also believed to make it possible for those asset items to possess better long-term performances. These two concepts, indeed, are the main goals of performance-based asset management contracts.

The scope of this research is limited to VDOT- issued contracts. This is mainly due to the fact that in developing the warranty clause template, special consideration was given to VDOT's needs and conditions and, most importantly, expectations from the contract. Nonetheless, some of the principles and approaches used during the development of the warranty clause template can also be referred to for the development of warranty clauses for the performance-based road maintenance contracts issued by the other state DOTs.

In developing the warranty clause template as a possible solution to address the subject of long-term performance, this research has not tried to elaborate on each issue that comes along with the warranty concept. Nonetheless, these issues, as described in the previous section, need to be considered before proceeding with the implementation of warranties. It could be concluded that that the warranty concept in highway contracting is a rather new concept which still needs a substantial amount of evaluation and which deserves much attention from interested parties.

Acknowledgments

This research has been funded by the Virginia Department of Transportation. The findings and conclusions are those of the authors and do not necessarily reflect the views of our sponsor. We would like to also thank the following individuals who provided input at certain stages of this research: Mr. Byron Blaschke, Mr. Mike Branch, Mr. Michael Hall, Mr. Stan Lanford, Mr. James Lowe, Dr. Jeffrey Russell, and Dr. Michael Vorster.

References

AASHTO Highway Subcommittee on Maintenance. (2002). *A guide for methods and procedures in contract maintenance.* AASHTO, Washington D.C., August.

Anderson, S.D., and Russell, J.S. (2001). "Guidelines for warranty, multi-parameter, and best value contracting." *National Cooperative Highway Research Program Report 451*, National Academy Press, Washington D.C.

Burrell, G., and Morgan, G. (1979). *Sociological paradigms and organizational analysis,* Heinemann, London.

Code of Virginia. (1995). "Public- Private transportation act."

Falls, L.C., Haas, R., McNeil, S., and Tighe, S. (2001). "Asset management and pavement management: Using common elements to maximize overall benefits." *Transportation Research Record 1769*. No: 01-2415.

FDOT. (2000). "Traffic stripes and markings- Performance based." *Section 707*. Feb. 16.
http://www.dot.state.fl.us/specificationsoffice/PerformanceBased/D7070000.d03.pdf

FDOT. (2002). "Contractor guaranteed asphalt pavement." *Section 338*. April 15.
http://www.dot.state.fl.us/specificationsoffice/PerformanceBased/D3380000IMP.pdf

Frost, M., and Lithgow, C.M. (1998). "Future trends in performance based contracting-Legal and technical perspectives." *IRR Conference*. Sydney, May.

Galehouse, L. (1998). "Innovative concepts for preventive maintenance." *Transportation Research Record 1627*. No: 98-0719.

Hancher, D.E. (1994). "Use of warranties in road construction." *NCHRP Synthesis of Highway Practice 195*, Transportation Research Board, Washington, D.C.

Hancher, D.E. (1999). "Contracting methods for highway construction." Transportation Research Board, Washington, D.C., November.

Krebs, S.W., Duckert, B., Schwandt, S., Volker, J., Brokaw, T., Shemwell, W., and Waelti, G. (2001). "Asphaltic pavement warranties, five year progress report." WisDOT, FHWA, WAPA, June.

MDOT (2002). "Transportation commission adopts pavement warranty policy." MDOT Press Release.http://www.michigan.gov/mdot/0,1607,7-151-9620_11057-31631--M_2002_4,00.html

ODOT. (2002). "Microsurfacing with warranty." *Supplemental Specification 881*. Oct. 18.
http://www.dot.state.oh.us/construction/OCA/Specs/SSandPN2002/8811002.pdf

ODOT. (2003). "Concrete pavement with warranty." *Supplemental Specification 884*.
http://www.dot.state.oh.us/construction/OCA/Specs/SSandPN2002/8840403.pdf

Ozbek, M.E. (2004). "Development of performance warranties for performance based road maintenance contracts." MS thesis, Civil Engineering, Virginia Polytechnic Institute and State University, Blacksburg.

Russell, J.S., Hanna A.S., Anderson, S.D., Wiseley, P.W., and Smith, R. J. (1999). "Current use of warranties in highway construction." Transportation Research Board Annual Meeting.

Stephens, J. Whelan M., and Johnson, D. (2002). "Use of performance based warranties on roadway construction projects." USDOT, FHWA, November.

VDOT. (1996). "Comprehensive agreement for interstate highway asset management services." December.

Appendix

ARTICLE XVI
WARRANTIES
WARRANTY FOR THE "_________" ASSET ITEM

Section 16.1: GENERAL DESCRIPTION

This document specifies the requirements for Contractor to warrant the maintenance work it (Contractor) performed for the "_________"[4] asset item for a period of _________[5] year(s)/month(s)[6].

Contractor shall assume full responsibility for all work specified in this Document for a period of _________[7] year(s)/month(s) after final acceptance of the Contract in accordance with **Section 11.3**[8], including continued responsibility (excluding the responsibilities as set forth in **Section 16.8** and **Section 16.10**) as to defects that are identified by the Department and as to which timely notice is provided within such warranty period until all such defects are corrected.

Section 16.2: WARRANTY BOND

Contractor shall provide a warranty bond for not less than the full warranty period, which commences upon final acceptance of the Contract in accordance with the **Section 11.3.** The effective starting date of the warranty bond shall be the date of final acceptance. The warranty bond will be released at the end of the warranty period or after all warranty remedial work has been completed, whichever is latest.

The surety of the bond shall be a surety company authorized to transact business in the Commonwealth of Virginia and shall have A.M. Best unit rating not less than "A"[9] throughout the warranty period. If the A.M. Best unit rating of the surety company drops below "A", then the Department shall have the right, which may be exercised at any time thereafter, but not the obligation to _________[10].

---

[4] Insert: the **name of the asset item** for which the warranty clause is developed.

[5] Insert: the **length of the warranty period** as identified (for the asset item for which the warranty clause is developed) using either Alternative 1 or Alternative 2 or Alternative 3.

[6] **year(s)** or **month(s)**, whichever is applicable.

[7] Insert: the **length of the warranty period** as identified (for the asset item for which the warranty clause is developed) using either Alternative 1 or Alternative 2 or Alternative 3.

[8] **Section 11.3** (which is under the Article XI of the current contract) establishes the completion of the Contract Term.

[9] This is the rating that has been required for the performance bond (which only covers the Contract Term) of the current contract.

[10] Insert: the **list of all of the steps that need to be taken** if the A.M. Best unit rating of the surety company drops below the required rating as set forth in this section.

The warranty bond(s) shall be in an amount not less than _________[11]. The bonds shall ensure the proper and prompt completion of required remedial work, including payments for all labor performed, and all equipment and materials used.

The warranty bond(s) shall be provided in one of the following manners:

1. A single term _________[12] year/month[13] warranty bond.

2. A _________[14] year/month renewable, non-cumulative warranty bond for _________[15] consecutive terms.

Regardless of the option chosen by Contractor, Contractor shall provide documentation evidencing proof of _________[16] year/month bond commitment before the execution of the contract.

No separate subcontractor bond is required. However, the bond(s) provided pursuant to this section shall cover all maintenance work, no matter whether such work is performed by Contractor or a subcontractor at any tier that is employed directly or indirectly by Contractor to perform a part or whole parts of the maintenance work during the Contract Term.

Section 16.3: WARRANTY COVERAGE AND WARRANTY CRITERION

Warranty criterion[17] is established by using the parameters listed in **Table 1**[18] below.

---

[11] Insert: the **bond amount** (for the asset item for which the warranty clause is developed) as established using the "most likely case" scenario.

[12] Insert: the **length of the warranty period** as identified (for the asset item for which the warranty clause is developed) using either Alternative 1 or Alternative 2 or Alternative 3.

[13] **year** or **month**, whichever is applicable.

[14] Insert: the **monthly or yearly** (whichever is applicable) **intervals** at which a non-cumulative warranty bond is required by VDOT.

[15] Insert: the **number** which is calculated by dividing the length of the warranty period as identified (for the asset item for which the warranty clause is developed) using either Alternative 1 or Alternative 2 or Alternative 3 to the monthly or yearly (whichever is applicable) intervals at which a non-cumulative warranty bond is required by VDOT.

[16] Insert: the **length of the warranty period** as identified (for the asset item for which the warranty clause is developed) using either Alternative 1 or Alternative 2 or Alternative 3.

[17] The warranty criterion is established using the performance criteria and the performance target for the asset item for which the warranty clause is developed. The warranty criterion for any asset item can be stated as "**meeting the performance criteria for at least equal or more than the performance target established for that asset item**".

[18] Insert: **Table 1** which lists the performance criteria and performance target (for the asset item for which the warranty clause is developed) that are established for the Contract Term within the Exhibit B of the current contract.

Table 1: Warranty Criterion

Asset	Outcome	Performance Target (% of asset that shall be in the outcome described condition)	Condition assessment acceptance tolerances and criteria (Performance Criteria)

This table needs to be extracted from the Exhibit B of the current contract. The information within this table needs to correspond to the asset item for which the warranty clause is developed.

Section 16.4: WARRANTY COMPLIANCE EVALUATIONS

The Department (and/or its authorized agent; referred to collectively hereafter in this Section 16.4 as "Department") will conduct an inspection at the beginning of the warranty period, which inspection shall be completed within _________[19]days from the warranty commencement date. For the remaining duration of the warranty period, the Department shall have the right, but not the obligation, to conduct additional periodic inspections between the warranty commencement date and the warranty expiration date. Notwithstanding the foregoing, in the event of a multi-year warranty, the Department will conduct at least one annual inspection during each year of the warranty term. The Department will conduct a final inspection within no more than _____ [20] days of the warranty termination date, which shall be completed within _________[21] days of the warranty termination date. The methodology of these inspections will be same as those conducted within the Evaluation Program that is defined in **Section 2.18**[22] and described in **Exhibit G**[23] of the Contract.

The Department will conduct these inspections at no cost to Contractor. The Department shall notify Contractor of the scheduled review. Contractor (and any applicable subcontractor at any tier, material supplier, or surety; referred to collectively hereafter in this Section 16.4 as "Contractor") may attend the inspections for observation purposes only. Any comments by Contractor may be recorded by the Department. A comprehensive report containing the results of the inspection and the remedial work (as expanded in **Section 16.5**) required from Contractor shall be made

[19] Insert: the **number of days** that is allowed for VDOT to complete the initial inspection. This time frame should be minimized to allow the Contractor to start remedial work as soon as possible.

[20] Insert: the **number of days** that is allowed for VDOT to begin the final inspection. This time frame should be minimized to allow the Contractor to start remedial work as soon as possible.

[21] Insert: the **number of days** that is allowed for VDOT to complete the final inspection. This time frame should be minimized to allow the Contractor to start remedial work as soon as possible.

[22] **Section 2.18** defines the highway maintenance evaluation program that is implemented in the Contract Term in order to assess the work of the contractor during the Contract Term.

[23] **Exhibit G** describes in detail the highway maintenance evaluation program that is implemented in the Contract Term in order to assess the work of the contractor during the Contract Term.

available to Contractor within __________[24] days after the completion of the inspection. The Department will endeavor to use only non-destructive procedures in performing the inspections. However, in the event the Department elects to employ destructive testing procedures, Contractor will not be responsible for damages that result solely therefrom.

The Department may also conduct special inspections on a more frequent basis in order to identify any emergency remedial work needs (as expanded in **Section 16.5**). If any emergency remedial work is identified during these special inspections, Contractor shall be notified within __________[25] hours upon the discovery of such required emergency remedial work.

As soon as the identified remedial work is completed (as expanded in **Section 16.5**), the Department shall begin and, within __________[26] days thereafter, complete a re-inspection of the item(s) for which the remedial work was required.

Section 16.5: REMEDIAL WORK

During the warranty period, if the results of the warranty compliance evaluations reveal that the warranty criterion (as expanded in **Section 16.3**) is not met, the remedial work, as identified to be necessary through the warranty compliance evaluations (that are expanded in **Section 16.4**), shall be performed by Contractor at no cost to the Department, subject to Contractor's right to challenge the Department's determination in accordance with **Section 16.6** hereof. At all times remedial work is being performed, Contractor shall follow a Department-approved traffic control plan.

Contractor shall , prior to commencement of such remedial work, submit a written plan for performing the required remedial work within __________[27] days after receipt of the Department's notification of required remedial work, except in case of emergency remedial work as detailed in this Section. Contractor's submittal need only to propose the schedule for performing the remedial work; no other information regarding Contractor's remedial work need to be submitted to the Department by Contractor. The times of the day for performing the remedial work shall be subject to the approval of the Department. The written plan shall be approved by the Department within __________[28] days upon the receipt of the

---

[24] Insert: the **number of days** that is allowed for VDOT to prepare and submit the report possessing the results of the inspection. This time frame should be minimized to allow the Contractor to start remedial work as soon as possible.

[25] Insert: the **number of hours** that is allowed for VDOT to notify the contractor of the emergency situation. This time frame should be minimized to allow the Contractor to start emergency remedial work as soon as possible.

[26] Insert: the **number of days** that is required for VDOT to complete the re-inspections.

[27] Insert: the **number of days** for the Contractor to submit its schedule to perform the required remedial work.

[28] Insert: the **number of days** for VDOT to approve the submitted schedule of the required remedial work. This time frame should be minimized to allow the Contractor to start remedial work as soon as possible.

written plan. Contractor shall commence the remedial work promptly upon the approval of its remedial work schedule.

The remedial work shall be performed by Contractor to improve the asset item's[29] overall condition sufficiently to meet the warranty criterion. The remedial work shall be completed within _________[30] days following the Department's approval of Contractor's remedial work schedule. If the remedial work cannot be completed within _________[31] days following the Department's approval of the Contractor's remedial work schedule due to seasonal limitations, Contractor shall notify the Department in writing and the Department and Contractor shall establish a mutually agreeable schedule for completion of the remedial work.

If Contractor does not agree the Department's reported inspection results, Contractor may initiate the Warranty Dispute Resolution Procedures as set forth in **Section 16.6**.

If, in the opinion of the Department, any portion of the highway covered by the warranty constitutes an emergency condition requiring immediate attention for the safety of the traveling public, Contractor shall be notified (as expanded in **Section 16.4**) to perform emergency remedial work. Conditions which necessitate emergency remedial work include, but may not be limited to _________[32]. If Contractor cannot perform the required emergency remedial work within _________[33] hours, the Department may have the work performed, at Contractor's expense, by the Department and/or its authorized agent. Any emergency remedial work performed by the Department and/or its authorized agent will not alter the requirements, responsibilities, or obligations of the warranty. If Contractor believes it should not be responsible for the performance and cost of any such emergency remedial work based upon its belief that the emergency condition was caused by factors beyond its control, Contractor may initiate the Warranty Dispute Resolution Procedures as set forth in **Section 16.6**.

If the remedial work necessitates the removal of asset items other than the one that is warranted under **Section 16.1** of this Contract, including, but not limited to, pavement markings, pavement markers, adjacent lane(s), roadway shoulders, signs,

---

[29] This is the asset item for which the warranty clause is developed.

[30] Insert: the **number of days** for the Contractor to complete all of the required remedial work related to the asset item for which the warranty clause is developed. This time frame should be minimized to restore the defective portions of the highway to their acceptable and safe conditions as soon as possible.

[31] Insert: the **number of days** for the Contractor to complete all of the required remedial work related to the asset item for which the warranty clause is developed.

[32] Insert: the **complete list of the occasions that constitute an emergency situation** for the asset item for which the warranty clause is developed.

[33] Insert: the **number of hours** for the Contractor to complete all of the required emergency remedial work related to the asset item for which the warranty clause is developed. This time frame should be minimized to prevent any safety issues that may arise due to these emergency situations.

and mile markers, then such asset items shall be removed and replaced at Contractor's expense.

At the end of the warranty period, Contractor will be released from further warranty work or responsibility, provided that all previously required remedial work has been satisfactorily completed.

Section 16.6: WARRANTY DISPUTE RESOLUTION PROCEDURES

A Warranty Dispute Resolution Board (the "Board") shall be established prior to the initiation of the warranty period to resolve any conflicts arising between Contractor and the Department regarding warranty requirements. The Board shall be composed of one representative appointed by Contractor, one representative appointed by the Department, and an independent party mutually agreed upon by Contractor and the Department. Decisions of the Board will be based on a simple majority vote. Contractor and the Department acknowledge that use of the Warranty Dispute Resolution Procedures is required and the determinations of the Board for conflicts arising between them will be binding on both Contractor and the Department, with no right to appeal by either party.

All Board meetings shall be initiated and paid for by Contractor. The Department will reimburse Contractor for all fees associated with any meeting of the Board only if the Board rules substantially in favor of Contractor; otherwise Contractor shall be solely responsible for all such costs. The term "substantially" is defined as greater than fifty percent of the issues when entitlement is disputed or greater than fifty percent of the total dollar amount when costs associated with such entitlement are disputed, or both.

The Warranty Dispute Resolution Procedures shall be initiated by Contractor within __________[34] days after the comprehensive report containing the results of the inspection and the remedial work required from Contractor is made available to Contractor (as expanded in **Section 16.4**). The Board shall meet within _____ [35] days after Contractor's initiation of the Warranty Dispute Resolution Procedures to

[34] Insert: the **number of days** for the Contractor to initiate the Warranty Dispute Resolution Procedures. This time frame should be minimized to allow the Board to begin evaluating the conflict as soon as possible so that the responsible party (as identified through the Warranty Dispute Resolution Procedures) can perform remedial work to restore the defective portions of the highway to their acceptable and safe conditions as soon as possible. Moreover, this time frame should be same as the time frame required for the Contractor to submit its schedule (as expanded in **Section 16.5**) for the remedial work. This would enable VDOT to understand the intentions of the Contractor in the sense that whether it is planning to perform the remedial work or it is planning to initiate Warranty Dispute Resolution Procedures for the requested remedial work.

[35] Insert: the **number of days** (after the initiation of the Warranty Dispute Resolution Procedures by Contractor) for the Board to meet in order to start the evaluation of the dispute. This time frame should be minimized to allow the Board to begin evaluating the conflict as soon as possible so that the responsible party (as identified through the Warranty Dispute Resolution Procedures) can perform remedial work to restore the defective portions of the highway to their acceptable and safe conditions as soon as possible.

evaluate the dispute. The Board's evaluation may include inspecting the disputed area in the field. The Board's evaluation may also include reviewing the inspection data, obtaining samples, and/or interviewing Department (District or Central Office) or Contractor employees. The Board shall receive the Maintenance Rating Program (MRP) training when it is determined necessary to perform an inspection to resolve the conflict. The Board's determination(s) shall be issued in writing and shall be provided to Contractor and the Department within __________[36] days after the Warranty Dispute Resolution Procedures are initiated by Contractor, unless otherwise mutually agreed by Contractor and the Department.

If the Warranty Dispute Resolution Procedures are initiated for an emergency remedial work (as expanded in **Section 16.5**) that has already been completed by the Department (and/or its authorized agent), then the Board shall determine whether Contractor was responsible for that emergency remedial work. As a result of this determination, the Board may direct Contractor to pay the Department for the completed emergency remedial work.

Section 16.7: DEFAULT ON CONTRACTOR

Failure of Contractor to submit its remedial work schedule within the specified period of time (as expanded in **Section 16.5**) shall be cause for the Department to complete the remedial work and recover the costs of such from the warranty bond.

Failure of Contractor to initiate the Warranty Dispute Resolution Procedures within the specified period of time (as expanded in **Section 16.6**) shall be cause for the Department to complete the remedial work and recover the costs of such from the warranty bond.

Failure of Contractor to perform remedial work within the specified period of time (as expanded in **Section 16.5**) shall be cause for the Department to complete the remedial work and recover the costs of such from the warranty bond.

Failure of Contractor to fully comply with the decisions of Warranty Dispute Resolution Board (as expanded in **Section 16.6**) shall be cause for the Department to complete the remedial work and recover the costs of such from the warranty bond.

Failure of Contractor or its surety to issue or renew the warranty bond (as expanded in **Section 16.2**) will be considered as a default and will result in forfeiture of __________[37] percent of the face amount of the bond to the Department.

Should Contractor fail to timely submit any conflict to the Warranty Dispute Resolution Board, fail to satisfactorily perform any remedial work, or fail to compensate the Department for any emergency remedial work performed by the Department, as determined by the Board to be Contractor's responsibility (as expanded in **Section 16.6**), the Department may suspend, revoke, or deny Contractor's certificate of qualification until the remedial work has been satisfactorily

[36] Insert: the **number of days** for the Board to arrive at a decision about the conflict(s) for which the Warranty Dispute Resolution Procedures were initiated. This time frame should be minimized to make the Board reach a final decision as soon as possible so that the responsible party (as identified through the Warranty Dispute Resolution Procedures) can perform remedial work to restore the defective portions of the highway to their acceptable and safe conditions as soon as possible.

[37] Insert: the **percentage** of the warranty bond which VDOT is willing to receive when declares default on the contractor for such issue.

performed or full and complete payment for the remedial work (and/or emergency remedial work) is made to the Department by Contractor or its surety. In no case shall the period of suspension, revocation, or denial of Contractor's certificate of qualification be less than __________[38] months.

Section 16.8: EXCLUSIONS AND WAIVERS

During the warranty period, Contractor will not be held responsible for defects that are caused by factors unrelated to Contractor's work. Factors which are considered to be beyond the control of Contractor are as follows: __________[39]. Other factors which are not listed here but which may be considered to be beyond the control of Contractor will be considered by the Department on a case by case basis upon receipt of a written request from Contractor.

Section 16.9: CONTRACTOR'S OPTION OF MAINTENANCE DURING THE WARRANTY PERIOD

The intent of the Contract is for Contractor to improve the __________[40] asset item's overall condition rating during the Contract Term in an effort to provide its long-term performance. It is also the intent of the Contract to provide a maintenance-free __________[41] asset item during the warranty period. Contractor may perform certain maintenance activities during the warranty period, but these maintenance activities are limited to the activities approved by the Department. To identify such maintenance activity needs, Contractor may monitor the highway using non-destructive procedures. Contractor shall not conduct any coring, milling, or other destructive procedures without prior approval by the Department.

Section 16.10: DEPARTMENT'S RESPONSIBILITIES

The Department (and/or its authorized agent) will be responsible for performing remedial work to correct the defects that are caused because of the factors that are beyond the control and without the fault or negligence of Contractor (as expanded in **Section 16.8**).

The Department (and/or its authorized agent) will be responsible for the incident management and winter weather maintenance activities. These activities can be listed as __________[42].

---

[38] Insert: the **number of months** as decided by VDOT to be the minimum length of the period in which the Contractor's certificate of qualification is suspended, revoked, or denied.

[39] Insert: the **complete list of the factors which are beyond the control of the contractor** and **which may cause defects** on the asset item for which the warranty clause is developed.

[40] Insert: the **name of the asset item** for which the warranty clause is developed

[41] Insert: the **name of the asset item** for which the warranty clause is developed

[42] Insert: the **complete list of the incident management** and **winter weather maintenance activities** that are applicable to the asset item for which the warranty clause is developed.

The Department (and/or its authorized agent) will be responsible for performing emergency remedial work immediately if Contractor is unable or refuses to perform such emergency remedial work (as expanded in **Section 16.5**).

The Department (and/or its authorized agent) will be responsible for monitoring the interstate highway system (that is maintained by Contractor during the Contract Term) to identify the remedial work and emergency remedial work needs (as expanded in **Section 16.4**) during the warranty period.

The Department will be responsible for submitting to Contractor the comprehensive report containing the results of the warranty compliance evaluations in a timely fashion (as expanded in **Section 16.4**).

The Department will be responsible for notifying Contractor in a timely fashion (as expanded in **Section 16.4**) of any remedial or emergency remedial work required to be performed by Contractor.

Section 16.11: SUBCONTRACTOR'S RESPONSIBILITIES

Contractor warrants any and all work, which includes the performance of its subcontractors at any tier. Upon receipt from VDOT of the notice for the remedial work (and/or emergency remedial work) that is needed for the maintenance work that was performed by a subcontractor at any tier, Contractor shall be responsible for enforcing or performing such remedial work (and/or emergency remedial work) in accordance with the Sections written under **Article XVI** of the Contract.

Miami Intermodal Center – Introducing "CM-At-Risk" to Transportation Construction

R. Edward Minchin Jr.[1], M.ASCE, Ketan Thakkar[2], and Ralph D. Ellis Jr.[3], M.ASCE

Abstract

The volume of traffic around the Miami International Airport Terminal has been a growing problem for many years. However, due to lack of funding and other issues, a project to relieve the congestion was never possible until now. The Miami Intermodal Center (MIC) will be a world-class multi-modal facility that is estimated to take 15 to 20 years to construct and cost approximately $2.25 billion.

MIC incorporates the first transportation project using the "CM-at-Risk" (CMAR) delivery system ever to receive federal funding. All of the work currently scheduled using the CMAR system is slated for Phase I of the project, which is now in progress and is estimated to take five years, costing $1.35 billion. The work to be done using the CMAR delivery system includes the rental car facility, the MIC terminal access roads, and other items, and is estimated to cost $230 to $250 million and be completed in 2007.

This paper describes the project, including the logistics of funding and how CMAR works. Since the early portions of Phase I have not been executed in a way to best utilize the strengths of CMAR, it is too early to tell how it is performing when compared to traditional project delivery systems.

Key Words: Miami Intermodal Center, CM-at-Risk, Delivery System, Access Facility, Terminal Access Roads, MIC/MIA, Tri-Rail

Introduction

Miami International Airport (MIA) is the first glimpse of the United States for many international travelers, especially those from Central and South America. It is also the first glimpse of Florida for many domestic travelers. To leave these travelers with a positive first impression of Florida and the United States is very important, but the appearance of the airport and the area around the airport has become less than desirable.

To make matters worse, traffic congestion around the terminal area of MIA was worsening by the month, stifling ground transportation. To mitigate this problem, several transportation construction projects were planned for the area, but

[1] Assistant Professor of Civil Engineering; University of Florida; 365 Weil Hall Gainesville, FL 32611; (352) 392-9537, Ext. 1488; Fax: 352-392-3394; minch@ufl.edu

[2] Research Assistant, University of Florida; 365 Weil Hall Gainesville, FL 32611; (352) 392-9537; Fax: 352-392-3394; ketant@ufl.edu

[3] Associate Professor of Civil Engineering; University of Florida; 365 Weil Hall Gainesville, FL 32611; (352) 392-9537, Ext. 1486; Fax: 352-392-3394; relli@ce.ufl.edu

even the most optimistic planners conceded that the roads around the terminal would become increasingly saturated over time even with the improvements planned.

Bold action was needed, and when funding became available, MIA, in conjunction with the Florida Department of Transportation (FDOT) and the Federal Highway Administration (FHWA), was finally able to act on the 1989 recommendation of the Miami International Airport Area Transportation Study for a multi-modal transportation facility adjacent to the airport.

Background

The project sponsor was FDOT, and because of its size and significance, the project was classified as a "mega-project" by the United States Department of Transportation. FDOT would act as the "owner" of the project even though FHWA, MIA, Miami–Dade County, and the Federal Aviation Administration (FAA) had a vested interest in the project. Once the decision was made to build the project, FDOT, already experiencing a quickly expanding construction and production program and having limited experience with the design and construction of vertical projects, was in the market for expertise and manpower to meet the challenges presented by the project. Therefore, a consultant, Earthtech, was quickly added to the team using standard FDOT procedures for consultant procurement. Earthtech has since acted as the Program Manager (PM).

Several decisions lay before the team. A designer had to be procured, as did construction engineering and inspection (CEI) services and construction services. A delivery system had to be chosen and numerous permits obtained. The proposed facility had to accommodate the rental car companies around the airport, the Tri-rail system (an above-ground train line that connects the three major south Florida counties—Dade, Broward, and Palm Beach—and ends at MIA), the Miami-Dade County Transit Authority (bus lines), Amtrack, Cruiseline Courtesy Buses, Hotel Courtesy Buses, Greyhound buses, the existing street and highway system, and numerous other transportation entities.

One of the most daunting tasks facing the team was how to please the numerous parties of interest that ranged from the other involved agencies, such the FAA and Miami-Dade County, to private interests, such as the Tri-rail system and the car rental companies that would use the new facility. In addition, any decision regarding the delivery system had to be filtered through FDOT's long-standing policy that any prime contractor has to self-perform at least 50% of the project.

Economic Impact

The construction of MIC will result in numerous temporary and permanent job creations within many sectors of the economy, especially construction, retail, and service. It is estimated that over the 15 to 20 years of the entire construction process, 76,000 construction and construction-related jobs will be generated, and 22,000 permanent jobs will be created to operate the facility once it is constructed. The economy and standard of living around MIA will be enhanced because the facility will encourage travelers to use the various public transportation systems integrated into the MIC. It is estimated that by 2010, 75,000 passengers will use the MIC daily. Of these, approximately 45,000 will be using the people-moving system, known as the MIA Mover, to travel to or from the airport (MIC, 2004).

Project Funding

Financing the MIC, with a total cost estimated at over $2.25 billion, was a challenge for the state of Florida and Miami-Dade County. Phase I is projected to cost approximately 1.35 billion over five years and has received funding from a variety of sources. For Phase I, the MIC has received nearly $165 million in Federal Highway Administration (FHWA) grants, $386 million in FDOT state funds, and a $25 million Florida State Infrastructure Bank (SIB) loan. The Miami-Dade Expressway Authority has provided $86 million in toll-backed funding, and the project has also received $25 million from Florida's SIB specifically for a portion of the project, the SR 836/SR 112 connector. The Miami-Dade Aviation Department will fund the $400 million MIA Mover with airport user fees.

Initially, the size of the MIC program prevented funding the entire project in a short period of time on a pay-as-you-go basis. The state of Florida, the Miami-Dade County Metropolitan Planning Organization (MPO), and Miami-Dade County had committed funding for Phase I of the overall project; however, the funds were spread over 15 years. These cash flow constraints would have caused the Phase I elements to be spread over 10 or more years, resulting in significantly higher costs for right-of-way acquisition and construction. In addition, this would have resulted in the disturbance of traffic in the area for an extended period of time.

The Transportation Equity Act for the 21st Century (TEA-21), passed by the U.S. Congress in June, 1998, contained a new federal credit assistance program originating from the Transportation Infrastructure Funding Innovative Finance Act (TIFIA), which was designed for better facilitation of financing for surface transportation projects of national or regional importance. TIFIA's principal goal is to use credit support rather than grants to help fast track large-scale projects. TIFIA provides credit support in three forms: loans, loan guarantees, and lines of credit. FDOT, in partnership with Miami-Dade County, applied for and received TIFIA assistance, and this funding has been influential in accelerating the MIC Phase I improvements.

Because of the two direct loans awarded under the TIFIA program in 1999 totaling $433 million, MIC was able to accelerate right-of-way acquisition and construction of the MIC Core, saving over $100 million in overall project cost. A $269 million TIFIA loan secured by state motor fuel tax revenues sped up the acquisition of right-of-way and initiation of work on the MIC core. The second TIFIA loan, for $164 million, was secured by rental car fees and was used to finance the consolidated RCF. Other major sources of funds include state gasoline tax revenues, SIB loans, and federal funds as described above.

Phases I and II

The transportation development portion of the project was separated into two phases. Phase I would include right-of-way acquisition, access improvements (roadways), the consolidated rental car facility (RCF), the MIA Mover, and commencing the MIC Core. Phase II, not scheduled to begin until at least 2007, would include completing the MIC Core, constructing platforms serving elevated Tri-rail, Amtrack, and Metrorail lines, and all construction of MIA landside facilities.

Phase I: While the entire program will be completed over the next 15-20 years, the first phase was scheduled to be completed over a five-year period of time at a cost of approximately $1.35 billion. This "Five Year Program" will consist of the following:

- Right-of-Way (ROW) Acquisition. Estimated Cost: $379 million
- Access Improvements (Roadways). Estimated Cost: $143 million
- Consolidated RCF. Estimated Cost: $162 million
- MIA Mover. Estimated Cost: $400 million
- MIC Core (Phase I). Estimated Cost: $80 million

Finance costs, contingency funds, and program management will add approximately $185 million to the cost of Phase I. The complete list of funding sources and their contribution can be seen in Table 1.

Rental Car Facility: The RCF will be the first major component of the MIC to be constructed. It will co-locate the rental car companies currently located at the airport and many of the companies located adjacent to the airport. The estimated cost of this facility is over $160 million. The sheer size of the building (1400 feet by 1200 feet, with the top floor sixty feet off the ground) demanded that everything possible be done to make the construction go as smoothly as possible; however, the size of the building was not the main consideration. Each floor of the building will house a fuel distribution center where gasoline may be pumped into the rental cars at each level.

Table 1. Funding Sources for MIC, Phase I

Source	Contribution ($Millions)
Prior and future Allocations of State and Federal Funds in Miami-Dade County's Transportation Improvement Plan (TIP), Long Range TIP, and Other State Funding	$249
MIA Capital Improvement Plan	$400
RCF Customer Facility Charge	$25
Miami-Dade Expressway Authority's Capital Program	$86
Miami-Dade County	$30
Ancillary Revenues for from Concessions and Joint Development	$37
FDOT State Infrastructure Bank Loan	$25
TIFIA Loan	$433
Capitalized Interest and Finance Costs	$64
Total	$1,349

This is the first time that any building in the United States has had elevated fuel distribution capability. Therefore special permits and numerous special considerations will be required. The RCF will include:

- Ready/return vehicle capacity of approximately 6500
- Fleet storage capacity (vehicles not in use) of approximately 3500
- Quick turnaround vehicle fueling and washing facilities
- Spacious customer service facilities for rental car transactions

MIC Core (Phase I).

The first phase of the MIC Core will cost approximately $80 million and accommodate the bus depot, Tri-rail passenger parking, and MIA Mover Station. The MIA Mover Station will be built adjacent to the RCF. This facility will allow passengers of Tri-rail, city busses, and rental cars to board a people-mover, known as the MIA Mover, and be transported to the airport terminal. The MIA Mover will feature large cars that run on either electric rails or rubber tires and will be boarded on the top floor of the building, 60 feet off the ground. This facility is expected to be completed about two years after completion of the RCF.

Delivery System.

Three delivery systems were given strong consideration for each aspect of Phase I; the result was that different portions of the work will be handled in different ways. Since federal funding was sought for each activity, FHWA approval was required for each delivery system decision. It was decided that portions of the work would be handled in the traditional FDOT way, which is by the linear method, or Design-Bid-Build (DBB). Other portions of the project would be handled by the Design-Build (DB) method.

For the RCF and the MIA Mover, a level of vertical construction expertise was needed that FDOT projects do not typically require. While to build these massive vertical facilities in a confined, urban environment presented a new challenge to FDOT, the earthwork and site preparation required were similar to that on FDOT projects. There were also roads and bridges called for in the project. Therefore, a delivery system was needed that would best meet the challenge of both linear and vertical construction.

In July 2000, Earthtech performed a technical evaluation of the three delivery systems as a means for the design and construction of the RCF and certain other related structures. The three systems considered were DBB, DB, and Construction Manager-at-Risk, or "CM-at-Risk" (CMAR). It was decided that this portion of work, amounting to approximately $230 to $250 million, would be let under a set of separate contracts, using the CMAR delivery system. Reasons given by Earthtech for this choice were that CMAR offered the following:

- Design process control
- Ability to meet or exceed schedule requirements

- Highly qualified contractor
- Highly qualified designer
- Budget/Cost control
- Project team formation
- Constructability input from the CM

A matrix used by Earthtech documenting the evaluation of the three candidate delivery systems is shown in Table 2. The matrix was designed by Sanvido and Konchar (1998) and customized and completed by Earthtech. Therefore, the rankings are the opinions of the raters and not necessarily those of the authors.

This choice in delivery systems meant that FDOT had to waive, for this project, their long-standing rule that all prime contractors had to self-perform at least 50% of project work. In addition, in order to receive federal funding, FDOT had to make a special application to FHWA under Special Experimental Project Number 14 (SEP-14). SEP-14 is a program by which FHWA can fund a limited number of projects using construction practices or delivery systems that they have no procedure or policy for accepting. Until fairly recently, a DB project had to go through this process, but after several successful DB projects, the FHWA stopped requiring this special application for DB projects. The MIC is the first CMAR project funded under SEP-14.

Table 2. Project Delivery System Evaluation Matrix

EVALUATION MATRIX RENTAL CAR FACILITY AND ASSOCIATED CONSTRUCTION	DESIGN-BID-BUILD	DESIGN-BUILD	CM@RISK	COMMENTS
ABILITY TO MEET SCHEDULE	X	●	●	SEE ATTACHED
UNIT COST EXPERIENCE	X	●	□	EMPIRICAL EXPERIENCE
QUALITY EXPERIENCE	X	●	●	EMPIRICAL EXPERIENCE
CONTROL OF CONTRACTOR SELECTION	X	●	●	
CONTROL OF DESIGNER SELECTION	●	●	●	
EARLY PROJECT TEAM SELECTION	X	●	●	
INTERACTION IN DESIGN PHASE	X	●	●	
EARLY CONSTRUCTIBILITY INPUT	□	●	●	
OPPORTUNITY TO PARTNER	X	●	●	
ABILITY TO PREQUALIFY PROJECT TEAM	N/A	●	●	
ABILITY TO PREQUALIFY SUBCONTRACTORS	□	□	●	
ABILITY TO OBTAIN RELEVANT EXPERIENCE	X	●	●	

● EXCELLENT
□ FAIR
X POOR
N/A NOT APPLICABLE

CM-at-Risk

Clough and Sears (1994) state in their well-known textbook that, "An appreciable share of the private construction market is now being done using the 'team approach.' When this procedure is followed, the private owner selects the architect and building contractor as soon as the project has been conceived. The three parties (then) constitute a team that serves to achieve budgeting, cost control, time scheduling, and project design in a cooperative manner."

Using the "team approach," the owner assembles his key players, such as the designer(s) and contractor(s), to study the proposed project. The team determines the project scope and budget, and the designer develops preliminary drawings from which the contractor makes conceptual cost estimates. As the process continues, the designer prepares the final drawings and specifications, and the owner makes the necessary financial arrangements. After financial commitments and required permits are obtained, actual construction begins. The designer and the contractor work closely together, modifying the design and the drawings as may be required. The process offers the owner the advantages of time savings, cost control, and improved quality (Clough and Sears, 1994).

The method chosen to facilitate fast-track construction of the MIC, CMAR, is very similar to this textbook description of the "team approach" and can be viewed as an adaptation of this established building construction delivery system used in private industry to civil or heavy construction in a public forum. The CMAR contract is, in essence, a cost-plus-a-fee contract with a guaranteed maximum price (GMP). It is the sum of the CM's fee, the CM's contingency, the General Conditions Construction, all of the subcontracts, and an estimate for unbid subcontracts. The CM agrees to pay for costs exceeding the GMP that are not a result of changes in the contract documents.

CMAR theoretically reduces the amount of risk for every entity involved in the project. From the beginning, with the client's (owner's) understanding of project requirements, the wisdom, experience, and technical expertise of architect-engineer (AE) and CM firms are combined. This team has control over every aspect of the project, and together they provide an absolute directive for design, construction, and functional requirements. One of the most important benefits is that the arrangement fosters a non-adversarial relationship that furthers collaboration in decision making. The CM can review the drawings beforehand and catch errors, reducing the owner's risk, while the AE similarly reviews the CM's approach to the work, providing constructive recommendations. The CM is allowed to take bids or proposals from subcontractors after entering the contract and prior to submission of the GMP, which reasonably reduces the CM's risk. The procedure is more methodical and more predictable than the low-bid DBB, and affords the owner more control over design than the DB system because the AE is under contract to the owner under CMAR instead of being part of a joint venture under contract to the owner and tied to the contractor, as is the case under DB.

The CM is expected to take care of any unexpected costs not caused by contract changes. Therefore, theoretically, the CM has built in a contingency within the GMP to cover unexpected but justifiable costs. The owner also has a contingency fund that is discussed later in the paper. When it comes to additional, unexpected costs incurred by subcontractors, as long as the subcontracts and work affected are due to changes in the contract documents, they are reimbursed to the CM. In essence then, any additional money paid to subcontractors for additional work (unless that work is from a change in the contract documents) comes straight out of the CM's profit, so the CM represents the owner in negotiating changes with subcontractors and is going to make very sure that the claims are legitimate.

CM-at-Risk at MIC

The MIC can be seen as a pilot project for CMAR from the FHWA standpoint because it is their first such project. It is also FDOT's first CMAR project but not the first for Florida. The Florida Department of Management Services (DMS) has built several vertical projects using the method over the previous 15 years. For this reason, DMS had personnel on the original Technical Review Committee and originally had a financial stake in the project because the CMAR contract was forged using DMS procedures. Immediately after the contracts were signed, DMS assigned all their rights to FDOT.

A negotiated fixed fee is the method of CM compensation on the MIC project. Originally, the CMs interested in this project had to submit a GMP for administering each segment of this fast-track construction project. These GMP were then subject to negotiation. The chosen CM is required to prequalify all subcontractors and oversee the bidding of all trades contracts. In this way, all construction work is competitively bid. The only way that the CM can self-perform any construction would be to outbid the subcontractors on a portion of the work. This has yet to happen.

The CM is paid a management fee plus expenses for oversight and coordination of the construction process. This includes project close-out, systems start-up, as-built drawings, operations and maintenance procedures, and warranty services. The CM also oversees the quality control (QC), quality assurance (QA), and Value Engineering (VE) for the project. Any savings realized by any VE change proposal are split, with 70% going to FDOT and 30% going to the CM. Earthtech was awarded the CEI work under a separate contract from the one for their Program Manager duties.

Construction items were broken down into separate GMP packages by FDOT, each to be negotiated separately and each with its own Letter of Authorization. The negotiations are between the CM (Turner) and the PM (Earthtech) acting on behalf of FDOT. The GMP packages are as follows:

- GMP I – RCF foundations and underground utilities. Estimated Cost: $17.6 million.
- GMP II – MIC Terminal Access Roadways (MTAR), tunnels, bridge. Estimated Cost: $25.2 million.
- GMP III – Tri-rail Station, MIA Mover Station, MIC/MIA Guideway Foundation. Estimated Cost: $45 million.
- GMP IV – RCF Building and Bridge. Estimated Cost: $160 million.
- GMP V – MIA Mover Lobby. No cost estimate at this time

On January 29, 2001, a legal notice advertisement requesting Statements of Qualification for a CM for this project was posted. Technical Proposals were submitted by the short-listed CMs on May 1, 2001, and oral presentations were heard by the Selection Committee on May 24, 2001. Turner Construction Company was chosen as CM for the project in a process similar to the selection process for a joint venture in a typical design-build project. Delays occurred in getting the project started due to the World Trade Center and Pentagon disasters of September 11, 2001, and again when the runner-up for the CM assignment protested the selection of Turner.

After the selection process, FDOT, DMS, and Turner Construction, Inc., signed a contract on March 1, 2003. Shortly thereafter, an architectural firm (Sequiera and Gavarette) was selected using established FDOT design professional procurement procedures. Personnel from Earthtech and FDOT served on the selection committee.

The GMP No. 1 bid package was issued to Turner on April 4, 2003, and the notice to proceed came in mid-July, 2003. The Organizational Chart for the MIC project is seen in Figure 1.

Phase II

The transportation development portion of Phase II will consist of completing the MIC Core, constructing platforms serving elevated Tri-rail, Amtrack, and Metrorail lines, and all construction of MIA landside facilities. There is also a commercial development portion of the MIC that will be completed as part of the Joint Development Agreement.

Joint Development Agreement.

The Joint Development Agreement was first conceived as a revenue-producing program and comprises the commercial development portion of the MIC. An oversight committee was formed to supervise the commercial development, and they retained ERA, a D.C.-based consultant, to perform a PDE study that found that the area in and around MIA was in need of office towers, parking, ancillary retail, and a hotel/conference center. In early 2002, an RFP was advertised for a developer to handle the commercial development, and the apparent choice is MIC Development, LLC, a joint venture consisting of equity partners The Codina Group and Mallory & Evans, hotel developer The Continental Company, marketing giant Market Place Development, and two large AE firms. Negotiations have begun with this group, with the expected start date on commercial development set for early 2008.

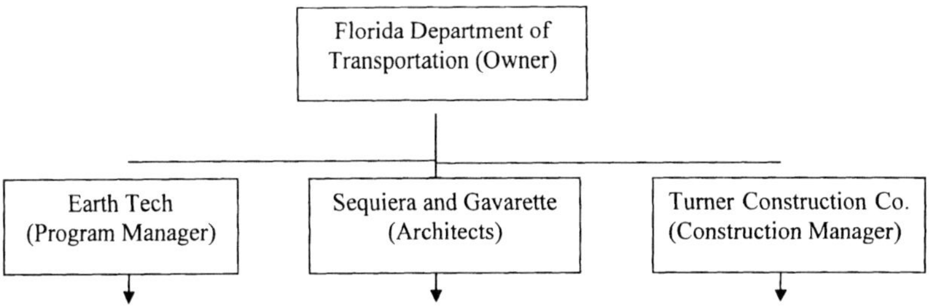

Figure 1. Organizational Chart for CMAR Portion of the Miami Intermodal Center

The Project

Project Progress.

As of the time of this paper submission, GMP No. 1 is 99% completed (one subcontractor has not had a portion of their work final accepted). To wit, approximately 2900 pre-stressed Portland cement concrete (PCC) piles (24 inches by 24 inches) had been driven, 584 pile caps had been constructed, 13,500 cy of grout had been placed, 9233 cy of PCC had been placed in pile caps, and 102,165 cy of fill dirt was delivered, spread, and compacted. GMP No. 2 is well underway, the notice to proceed for GMP No. 3 has been issued, and GMP No. 4 is currently being negotiated by Turner and Earthtech..

System Evaluation.

The CMAR delivery system is supposed to bring a wide variety of advantages to a construction project; however, there are important elements that must be in place if CMAR is to bring all of its potential benefits to the project. Two of these essential elements are: 1) The CM must be on board no later than the 35% plans review (preferably earlier), and 2) There must be a designer that works especially quickly and efficiently between the 65% plans review and plans completion.

If the CM is not on board by the 35% plans review, then the construction expertise that is so much a part of the CMAR design phase will be lost. In addition, if the designer does not move quickly from the 65% plans review to plans completion, the fast-track advantage will be lost because the CM cannot obtain many of the required permits without completed plans.

Unfortunately, Turner was not placed in a working relationship with the AE for GMP No. 1 until the plans were completed; therefore, there was not the level of constructability review necessary for an ideal job. In addition, the project was perilously close to being severely delayed by permit acquisition. In both cases, FDOT and the rest of the project team were fortunate. Because of the nature of the changes, although 50 change orders were executed in GMP No. 1, resulting in a net cost increase of approximately $600,000, very few of them, according to the PM, could have been avoided by a proper constructability review. As for the permits, the project team was again fortunate. The county allowed the RCF to be built under the old (2000) South Florida Building Code instead of insisting it be built under the new (2002), more demanding and time-consuming Florida Building Code, which saved time and outlay equaling at least $5 million.

Another big advantage of CMAR is that, theoretically, there should be a better relationship between the parties to the contract than under the DBB system. While all agreed that this was the case during GMP No. 1, it is apparent that not having the CM on board until plans completion caused things to occur on the project that strained relations to some extent. For example, the CM was forced to act "more like a prime contractor than a CM" in some of the dealings with the subcontractors because of constructability issues and other questions that could have been solved proactively instead of reactively had the CM been on the team from the beginning. There were also issues between the PM and the CM early in the process of GMP No. 1 that could have been avoided by having the whole team in place from the beginning. (MIC Project Personnel 2004)

Even with the CM coming on board late in GMP Nos. 1 and 2, the evaluation of the system by the participants was positive overall. Project personnel reported a "great" relationship between the CM and their subcontractors and that there was much less of an adversarial relationship between all parties to the contract than has been experienced under the DBB system.

The difference between working on a project under the DBB system and under the CMAR system was described by one individual on the project as "two different planets. Under the DBB (low-bid) system, all of your profit is made on project changes. Contractor or CM personnel are trained to find potential changes." With CMAR, the CM is responsible for the whole job and the approach to the project is different. The CM under CMAR is much more service-oriented than a CM or prime contractor under DBB. In fact, when this company hires someone out of the DBB environment, it reportedly takes an individual about two years to convert from the adversarial attitude bred by the DBB delivery system to the more service-oriented attitude prevalent in the CMAR delivery system. (MIC Personnel, 2004)

Other experiences and insights shared by those involved with the MIC project include:

- The key element in the CMAR system on this project is the contingency fund (10% on this project). The PM has a sum of money from which they may, at their discretion, grant the CM relief for costs incurred from additional work resulting from changes in the contract documents. According to project personnel, without the contingency fund, an adversarial atmosphere would likely appear on the project.
- Instead of the prime contractor or the CM looking for changes as on a DBB project, the subcontractors are doing so, but a strong CM insulates the owner from this problem.
- There was freedom to attract a blue chip contractor (CM), not the low bidder.
- There is much more flexibility and ability to handle the unexpected.

- There is a level of control over the design process that is not possible within an arrangement where the designer and the constructor are contractually linked.
- Much greater ability to handle things that the owner and CM are not familiar with, such as the elevated fueling facilities (MIC Personnel 2004).

Fortunately, for GMP No. 2 the CM was brought in a little earlier than in GMP No. 1; however, it was not early enough. Turner was brought in only a little before the plans were completed but was able to influence the design via the VE process even though they were unable to help at all with the constructability review function. In short, in GMP No. 2 Turner has been more involved in the design than in GMP No. 1 but still not fully utilized. This will undoubtedly result in some problems that could have been avoided if the CM had been on the team from the beginning as prescribed for CMAR. Thankfully, for GMP Nos. 3 and 4 Turner was brought in at the beginning of the design phase under a constructability contract separate from the GMP contract. GMP Nos. 3 and 4 will allow a much better assessment of the potential of the CMAR delivery system for transportation construction.

For GMP Nos. 2, 3 and 4, Turner's estimate far exceeded FDOT's budget, so negotiations were used to reconcile these price differences. In addition, Turner conducted VE studies to further reduce costs in order to attempt to meet FDOT's original budget. Early indications from negotiations for GMP No. 4 indicate that FDOT may have to pay higher premium and contingency costs but will share in the savings that the CM (Turner) can negotiate for subcontracted work. Here again, the savings splits as 70% to FDOT and 30% to Turner.

FDOT and the PM are taking the project one GMP package at a time, and due to Turner's submitted price for GMP No. 4 being well over the budget, GMP No. 4 may be let for DBB or to another CM if the differences cannot be resolved. If this happens, then GMP No. 5 would still be let CMAR if the GMP submitted by Turner or another CM allows for it. FDOT has total flexibility and authority to determine whether to proceed with CMAR or switch to another delivery system before the notice to proceed is issued for any GMP.

A very interesting part of the project will be the construction of the elevated fueling centers. The Senior Program Director for the project said that for the period of time that this portion of the RCF is being constructed, "the Fire Marshall will be running the project" (MIC Personnel 2004). This unique construction will be a subject of widespread interest.

Summary and Conclusions

The MIC is a recognized mega-project and a will be a significant component of south Florida's transportation network. It will help solve mobility problems that plague the area and will promote goals outlined in TEA21 legislation while enhancing the long-term viability of MIA.

A TIFIA loan has played a critical role in the overall financial structure of the MIC program. It has allowed the parties to deliver the project in the most efficient and timely manner. Winning the TIFIA loan approval won world-wide recognition for MIC.

Although the CM was not involved as early as needed in GMP Nos. 1 or 2, the CMAR system has still performed in a way that is satisfactory to the owner. The adversarial relationship between owner and CM or prime contractor under the DBB system was greatly reduced even in this incomplete execution of the system. Many of the other advantages credited to the CMAR system were also apparent in the execution of GMP No. 1. Among these are:

- With CM-at-Risk, FDOT has a level of control over the design process that is impossible within any arrangement wherein the designer and the constructor are contractually linked and the owner has a contract with each only as part of a joint venture. Similarly, for the construction work, it provides an opportunity to select a contactor (CM) with the expertise needed for a particular project and an excellent track record of completing projects on time and within budget.
- Instead of the prime contractor or the CM looking for changes to increase profits, the subcontractors are doing so, but a strong CM insulates the owner from this problem.
- There is much more flexibility and ability to handle the unexpected

For GMP No. 2, the design was also almost at 100% when Turner was brought in; thus, no constructability reviews were provided by Turner. This time, however, Turner did make some contribution to the pre-construction phase by providing VE for some portions of the work.

CMAR was used as planned on GMP, Nos. 3, and 4 (and hopefully 5). This is more true to the theoretical CMAR because Turner will be thoroughly involved in the design and construction activities with input from the beginning through constructability reviews and VE. Therefore, GMP Nos. 3, 4, and 5, if CMAR is used on each, will provide a better case study of the delivery system than the first two GMPs. In order for CMAR to function properly, the CM must be involved very early in the process. All project personnel interviewed agreed that they were very fortunate that GMP 1 worked out as well as it did, and they don't intend to rely on good fortune in future GMPs.

Much of the credit for the flexibility and service orientation of the CM and subcontractors was directed at the contingency fund (10% on this project). Without that, it was feared that an adversarial relationship would manifest itself on the project.

FDOT is very pleased, at this point, with its decision to use the CMAR delivery system, and the research team will continue to monitor the project to learn more about the application of this method to transportation construction.

References

American Association of State Highway & Transportation Officials. (2004). Innovative Finance for Surface Transportation, http://www.innovativefinance.org

Bai, Yong and Hezam, Adel A. (2003). *Integrating Innovative Project Delivery Methods into the Construction Curriculum,* ASC Proceedings of the 39th Annual Conference pp 119-128.

Barrie, Donald S. & Paulson, Boyd C. (1992). *Professional Construction Management* (2nd Ed.). McGraw-Hill, Inc., New York.

Berman, H.W. (1999). *Understanding Project Delivery Methods*, ALI-ABA's The Practical Real Estate Lawyer, Volume 15, Number 2.

Clough, Richard H. & Sears, Glenn A. (1994). *Construction contracting* (6th Ed.). John Wiley & Sons. Inc, New York.

Dorsey, Robert W. (1997). *Project delivery systems for building construction.* Associated General Contractors of America, pp. XI.

Florida Department of Transportation. 2004. United States Department of Transportation http://www.dot.state.fl.us

Miami Intermodal Center (MIC). 2004. Florida Department of Transportation, http://www.micdot.com

Florida Department of Transportation, District VI, Miami Intermodal Center (MIC) News letter. (Fall/Winter 2000).

MIC Project Personnel (2004). Interview with several, including personnel with FDOT, Earth Tech, Turner, and Serianni.

Sanvido, Victor, and Konchar, Mark D. (1998) *Project delivery Systems: CM at Risk, Design-Build, Design-Bid-Build.* Construction Industry Institute

Sumoski, James A. P.E., (2004). MIC Construction Project Manager, State of Florida, Department of Transportation.

Implementing Best-Value Procurement in Highway Construction Projects

By Douglas D. Gransberg, PE, M.ASCE,[1] Keith R. Molenaar, AM.ASCE,[2] Sidney Scott, PE, M.ASCE,[3] and Nancy Smith[4]

Abstract

In today's construction climate, state highway agencies are finding themselves under increasing pressure to improve project performance, complete projects faster, and reduce the cost of administering their construction programs. In response to these pressures, the industry is turning to alternative contracting methods such as best-value performance to assure project quality and enhance performance. In essence, best-value procurement incorporates construction contractor selection process factors, which along with price, seek to improve project performance or achieve other specific project goals. The paper reports the results of the analysis of the procurement documents from over 50 best-value projects. The contents of the best-value selection plans are categorized into best-value parameters, evaluation criteria, scoring systems and award algorithms. The paper also summarizes regulatory trends and survey results from state highway agencies concerning best value use. A synthesis of practices currently in use in numerous public construction sectors leads to the conclusion that best-value contracting processes are appropriate to select construction contractors as part of the traditional design-bid-build delivery process for highway construction.

Introduction

This paper examines the state of the practice in best-value procurement methods in the construction industry found in the literature, project procurement documents, and survey data. It includes regulatory trends, concepts found in the literature and project data, parameters used in the process, and case study information from both design-bid-build (DBB) and design-build (DB) projects to illustrate how best-value has been implemented.

The development of best-value procurement concepts in the public sector has to some extent borrowed ideas and approaches used to procure products and services in the private sector. Private sector construction owners have long sought to get the "best-value" for dollars expended. For example, Proctor and Gamble, a major U.S. corporation with an annual construction budget of $1.5 billion, has often used best-

[1] Professor, Construction Science, University of Oklahoma, Norman, Oklahoma 73019, dgransberg@ou.edu, (405) 326-6092.

[2] Associate Professor, Civil, Environmental and Architectural Engineering, University of Colorado, Boulder, Colorado 80309, keith.molenaar@colorado.edu, tel (303) 735-4276, fax (303) 492-7317.

[3] Vice President, Trauner Consulting Services, Inc., Philadelphia, Pennsylvannia 19103, sid.scott@traunerconsulting.com, 215-814-6400.

[4] Partner, Nossaman, Guthner, Knox & Elliot, LLP, Los Angeles, California, 90071, Nsmith@nossaman.com, 213-612-7837.

value selection with a negotiated procurement for industrial projects. Contractor selection is typically based on multiple factors that include cost, schedule, quality management, safety, and technical ability (Dorsey, 1995). Best-value procurement practices are increasingly being transferred to the public sector where permitted by legislation or when determined to be in the best interests of the agency under both traditional and alternative contracts. Although legislative requirements have traditionally required low bid for construction, there are more and more examples of revisions to legislation allowing best-value procurement.

Legislative and Regulatory Trends

Legislation and regulations for public sector construction at the federal and state levels are moving towards greater use of contracting approaches to achieve the "best-value" for dollars expended. In some cases, particularly at the state level, the statutes specifically address the use of best-value selection in conjunction with competitive sealed bidding. The Federal Acquisition Regulation (FAR), Part 15 – Contracting By Negotiation, sets forth best-value concepts under a competitive acquisition. (FAR 2000) Best-value under the source selection process might entail the selection of the lowest-priced technically acceptable proposals or it may consist of a trade-off between price and other factors – resulting in the section of a proposal without the lowest price.

Many federal and state agencies have implemented various source selection methods and have developed instructions or procedures for development and implementation of these methods. At the Federal level, the U.S. Postal Service, the Army, the Navy, the Department of Veterans Affairs, and the Federal Bureau of Prisons have developed procedures and guidelines for source selection contracting applicable to their construction programs. (U.S. Postal Service 2000; U.S. Department of Justice 2000; Army 2001) Though federal legislation has not explicitly directed the use of best-value for highway construction, the Federal Highway Administration (FHWA) has for many years allowed alternative procurements using best-value concepts embedded in trial or experimental contracting methods for selected highway projects through its Special Experimental Project (SEP-14) initiative. The lessons-learned from this program have added to the body of knowledge for best-value procurement in the highway sector (FHWA 1998).

At the state level, various statutes have addressed best-value under both competitive sealed bidding and alternative contracting for public works contracts. Statutes addressing best-value in the context of competitive bidding are of particular interest for this research. The Colorado Revised Statutes includes rules for "competitive sealed best-value bidding." Excerpts from the statute indicate that, when determined to be advantageous to the state, the procurement officer may "...allow bidders to submit prices for enhancements, options, or alternatives that will result in a product or service to the state having the best-value at the lowest cost." The statute further allows that a contract may be awarded to a bidder where the total amount of a bid price and the prices for enhancements, options, or alternatives of the bidder exceed the total amount of the bid price and the prices for enhancements, options, or alternatives of another bidder if it is determined that the higher total amount provides a contract with the best-value at the lowest cost to the state (Colorado 2002).

The Delaware Code similarly allows the use of best-value procurement for large public works contracts in conjunction with a determination of responsiveness through pre-qualification. In Delaware if an agency elects to award based on best-value, the agency must first determine that the successful bidder is responsive and responsible. The determination of best-value is then based upon objective criteria that have been communicated to the bidders in the invitation to bid. Objective criteria are assigned a weight consistent with the following:

- Price --- must be at least 70% but no more than 90%;
- Schedule --- must be at least 10% but no more than 30%;

A weighted average stated in the invitation to bid is then applied to each criterion according to its importance to each project. The agency ranks the bidder according to the established criteria and awards to the highest ranked bidder (Delaware 2001).

Similarly, excerpts from the revised Kentucky Revised Statutes define best-value as: "Best-value means a procurement in which the decision is based on the primary objective of meeting the specific business requirements and best interests of the Commonwealth" (Kentucky 2001). These decisions must be based on objective and quantifiable criteria that includes price and have been communicated to the offerors in the invitation for bids.

Thus, legislation at the federal and state levels is allowing the use of best-value selection strategies that include price and other factors when deemed to be in the best interests of the agency. One can see that legislative bodies are allowing public highway agencies more flexibility in the procurement of typical projects. As can be seen in Delaware, legislators are also often putting objective limits on how far an agency can move from the traditional low bid by assigning a relative weight between the price and other factors. This should probably be viewed as a part of the learning curve that must take place as the public works low bid procurement culture is replaced by a best-value procurement culture. It also serves to illustrate the willingness of elected bodies to allow their subordinate agencies to experiment with new procurement methodologies. Finally, it shows that the best-value concepts, analysis, and recommendations presented in this research fall within the framework of current federal and state legislation.

Research Methodology

The research methodology consisted of coordinating the results of three main actions. First, a comprehensive literature review of procurement methods used in the construction industry within the past 15 years was undertaken to identify the current state of best-value contracting. Many of the findings of this review highlight issues and shortcomings in the traditional low bid system and address trends in public sector construction towards the increased use of various best-value procurement methods to improve project performance and enhance end-product quality. The literature draws from all facets of the construction industry in the United States, Europe, and other countries. It includes perspectives from federal and state contracting agencies, vertical and horizontal construction, and analysis of project outcomes correlated to various procurement systems incorporating non-cost factors in the selection process.

Next, a content analysis was conducted on 50 best-value procurement documents obtained from 23 state departments of transportation (DOT) and 15 federal agencies, as well as other public agencies at the state, local, and international level. The documents were reviewed and their contents inventoried in accordance with the best value concepts shown in the next section of this paper. The purpose of this analysis was to identify how public agencies are implementing best-value contracting principles in their actual procurement documents and compare that to the information found in the literature review and in the next piece of the study which is the survey of transportation professionals.

Finally, a questionnaire was designed to identify the current state of practice in the industry and to identify key respondents that could provide additional project-related information for follow-up case studies. It identified transportation agencies that are using or considering the use of a best-value procurement process consistent with the definitions and concepts discussed in the previous section. It also asked respondents to identify any new best-value concepts that may not be reflected in the literature or our existing database. The survey was e-mailed to American Association of State Highway and Transportation Officials (AASHTO) representatives from each of the 52 AASHTO agencies and various other affiliated transportation organizations. The initial contact list consisted of representatives from the AASHTO Subcommittee on Construction and related highway organizations. The survey asked that questions should be completed by the personnel responsible for procuring and administering the agency's construction program, particularly with regard to alternative contracting methods. 44 responses from 41 transportation agencies were received. Finally, the results of the three areas of study were then synthesized to permit the development of conclusions. Those were organized in the same format as the best value concepts.

Best-Value Contracting Concepts

In a broad sense, the definition of best-value may encompass the concepts from and variations of current highway procurement methods contained in the FHWA SEP-14 program, including pre-qualification, post-qualification, A+B bidding, multi-parameter bidding, bid alternates, and extended warranties. While this definition is all-inclusive, four primary concepts within best-value procurement must be discussed to refine the definition and make it applicable to highway construction procurement. These concepts are illustrated in Figure 1. The research team conducted a series of over 50 case studies to illustrate the application of the best-value concepts. These projects came from a wide variety of public agencies and included both transportation and building projects. Fourteen of these projects that were examined in detail are shown in Table 1, which is provided to graphically illustrate the how the research methodology is applied to the case study procurement document content analysis to arrive at a uniform structure with which to apply the best-value concepts shown in Figure 1 to any given project. A complete table of case studies is presented in a summary table in Appendix 1.

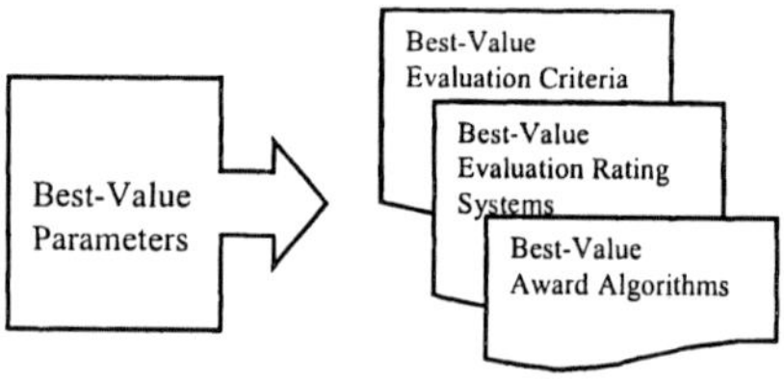

Figure 1. - Best-Value Concepts

Table 1. Case Study Summary Table

Case	Parameters	Award Algorithm	Eval Rating System
1. Air Force Base Pedestrian Bridge	A.0 + P.1	Qualitative Cost-Technical Tradeoff	Adjectival Rating
2. NASA Johnson Space Center Tunnel System	A.0 + P.0 + P.1	Qualitative Cost-Technical Tradeoff	Adjectival Rating
3. Corps of Engineers Canal	A.0 + P.1 + P.2 + P.4	Qualitative Cost-Technical Tradeoff	Not stated
4. Swedish Highway Administration Asphalt Paving Bids	A.0 + P.1 + P.2 + P.4 + D.0	Weighted Criteria	Direct Point Scoring
5. Alaska DOT Interchange	A.0 + A.1 + P.0 + P4 + D1	Weighted Criteria	Direct Point Scoring
6. University of Nebraska Cleanroom	B.0 + P.0 + P.2 + P.4 + D.1	Fixed Price – Best Proposal	Direct Point Scoring
7. US Army Corps of Engineers Dam	A.0 + B.0 + P.1 + P.2 + P.3 + P.4	Qualitative Cost-Technical Tradeoff	Satisficing and Adjectival Rating
8. Spanish Road Association Asphaltic Paving and Highway Maintenance	A.0 + B.0 + P.1 + P.2 + P.3 + P.4	Weighted Criteria	Direct Point Scoring
9. Minnesota DOT Highway	A.0 + B.0 + P.0 + P.1 + Q.0 + D.1	Meets Technical Criteria – Low Bid	Satisficing
10. Missouri DOT Bridge Seismic Isolation System	A.0 + A.1 + B.0 + P.1 + P.3 + Q.0 + D.0	Meets Technical Criteria – Low Bid	Satisficing
11. Washington State DOT Interchange	A.0 + B.0 + B.2 + P.0 + P.1 + P.2 + P.4 + Q.0 + Q.4	Adjusted Score	Direct Point Scoring
12. US Army Corps Air Freight Terminal/Airfield	A.0 + B.0 + P.1 + P.2 + P.3 + P.4 + Q.0 + Q.4 + D.0	Meets Technical Criteria – Low Bid	Modified Satisficing
13. US Forest Service Highway	A.0 + B.0 + B.2 + P.0 + P.1 + P.2 + P.3 + P.4 + Q.4 + D.1	Quantitative Cost-Technical Tradeoff	Direct Point Scoring
14. Maine DOT Bridge	A.0 + A.1 + B.0 + B.2 + P.0 + P.4 + Q.0 + Q.2 + Q.3 + Q.4 + D.1	Adjusted Bid	Direct Point Scoring

The following sections describe each of the four best-value concepts in detail. The discussion is based on a review of the current literature and examination of actual best-value solicitation documents from all section of public construction and purchasing.

Best-Value Parameters

From inspection of the literature, the research team determined a number of best-value parameters that can be mixed and matched to create a best-value highway contract. These parameters define individual elements, which can be combined to create an appropriate best-value definition, evaluation, and award. Many of these parameters overlap with multi-parameter bidding practices, but the parameters described herein are much more comprehensive than those described in previous NCHRP multi-parameter contracting literature (NCHRP 2001). Each parameter has been given an alphanumeric identification to make classifying each case study procurement document easier for the reader to understand, and each flows out of a combination of five major areas as shown below:

- Cost: This area includes any and all items that are expressed in financial terms.
    - Bid Cost = A.0
    - Life-cycle Costs = A.1
- Time: This area covers all items that either define or affect the project's schedule.
    - Schedule = B.0
    - Lane Rental = B.1
    - Traffic Control = B.2
- Qualifications/Performance: This area includes both the personal qualifications and experiences of key project personnel and the corporate experience of the entities that will complete the construction
    - Prequalification = P.0
    - Past Project Performance = P.1
    - Key Personnel Experience = P.2
    - Subcontractors Information = P.3
    - Project Management Plans = P.4
- Quality: This area covers all aspects of the project's quality management system including mechanisms to ensure long-term quality such as warranties. Most are not biddable in that they cannot be converted to a dollar or time amount.
    - Warranty = Q.0
    - Warranty Credit = Q.1

    - Quality Parameter measured with % in limits = Q.2
    - Quality Parameter using performance indicator = Q.3
    - Quality Management Plans = Q.4

- Design Alternates: This area covers any aspects of the project's design that the construction contractor is allowed to either propose or alter including the entire design as in a design-build project:
    - Design with Bid Alternate = D.0
    - Design-Build with Performance Specifications = D.1

Thus, we can create a set of potential variations on the theme of best-value that is equal to the number of statistical combinations that we can develop using two or more of the above. For example, a best-value project that has cost, schedule, prequalification, past project performance, and a quality parameter using performance indicator would be: A.0 + B.0 + P.0 + P.1 + Q.3

Best-Value Evaluation Criteria

Upon definition of the best-value parameters for a project, the agency must embark upon creating an evaluation and award plan. This evaluation plan will involve creating best-value evaluation criteria from the previously mentioned parameters, defining evaluation criteria rating systems, and finally defining a best-value award algorithm.

Best-value evaluation criteria include those factors, in addition to price, that add value to the procurement. Evaluation criteria vary on each project. In addition to the detailed case studies, the best-value evaluation criteria were summarized from 50 requests for proposals (RFPs) as shown in Table 2. Each is classified into one of the best-value parameters previously discussed. It illustrates the additional information gleaned from the analysis of best-value RFP's collected during this study.

The solicitation documents included both vertical (building) projects and horizontal (transportation/utility) projects. The population concentrated on design-bid-build/best-value RFP's specifically, but as best-value contracting is in it's infancy in highway construction, the population also looked at design-build projects in highways to find those types of evaluation criteria used that would easily be translated to design-bid-build/best-value contracts. The vertical projects were surveyed for the same reason. It can be seen that most of the criteria can be fit into one of the best-value parameter definitions. Public agencies also must include regulatory evaluation criteria to comply with their local procurement law constraints.

Table 2. Summary Evaluation Criteria as Identified with Best-Value Parameter from Total Case Study Project Population

Evaluation Criteria (1)	Best-Value Parameter	Number of Contracts Using Evaluation Criteria (Total = 50) (2)
Price Evaluation	A.0	42
Low Bid	A.0	7
Life-cycle Cost	A.1	2
Project Schedule Evaluation	B.0	19
Traffic Maintenance	B.2	3
Financial & Bonding Requirements	P.0	35
Past Experience Evaluation	P.1	44
Safety Record (or plan)	P.1	25
Current Project Workload	P.1	17
Regional Performance Capacity (political)	P.1	4
Key Personnel & Qualifications	P.2	41
Utilization of Small Business	P.3	30
Subcontractor Evaluation/Plan	P.3	29
Management/Organization Plan	P.4	31
Construction Warranties	Q.0	11
Construction Engineering Inspection	Q.2	1
Construction Methods	Q.3	1
Quality Management	Q.4	27
Proposed Design Alternate & Experience	D.0	26
Mix Designs & Alternates	D.0	2
Technical Proposal Responsiveness	D.1	37
Environmental Protection/Considerations	D.1	25
Site Plan	D.1	5
Innovation & Aesthetics	D.1	5
Site Utilities Plan	D.1	1
Coordination	D.1	1
Cultural Sensitivity	D.1	1

Examination of Table 2 provides that Cost and Qualifications criteria are the most heavily used in all types of best-value contracts. Past performance, qualifications of key personnel, and subcontracting/small business plans are the most popular of the Qualifications parameter criteria. In the Quality parameter group, evaluation criteria for quality management planning and warranties led the category. In the Design parameter, criteria specifying an evaluation of technical proposals were used in the majority of the RFP's. The heavy use of this criterion must be jointly compared with the use of the "proposed design alternates" criterion to gain a feeling for the amount of design detail the agencies were willing to allow the contractor to apply to the project. Fifteen of the case study projects were design-build projects, and all of those would be required to propose a design alternate that would need to be evaluated. Therefore, there were 11 design-bid-build projects in the population where the agency asked for a design alternate for evaluation. Proposed environmental protection measures were also a popular aspect of design information that public agencies wanted to evaluate. Finally, among the design-related evaluation criteria, proposal responsiveness was the preeminent criterion as would be expected.

Best-Value Evaluation Rating Systems

Public owners have used a variety of evaluation (scoring or rating) systems for the evaluation criteria discussed in the previous section. Many evaluation rating systems are quite simple, while others are quite sophisticated. Different best-value parameter evaluation criteria may require unique evaluation rating systems. All can be generically categorized into to the following four general types of systems whose details are shown in Table 3:

- Satisficing (more commonly called "Go/No Go"): Satisficing is by the simplest and easiest to understand for both evaluators and proposers. To use it, the evaluator must establish a minimum standard for each and every evaluation criterion against which the proposals can be measured. This is relatively easy for certain kinds of criteria such as qualifications.
- Modified Satisficing: Modified Satisficing recognizes that there may be degrees of responsiveness to any given criteria. As a result, the range of possible ratings is expanded to allow an evaluator to rate a given category of a proposal across a variety of degrees. Thus, a proposal that is nearly responsive can be rated accordingly and not dropped from the competition due to a minor deficiency. Additionally, a proposal that exceeds the published criteria can be rewarded by a score that indicates that it exceeded the standard. Modified satisfied systems usually differentiate between minor deficiencies that do not eliminate the bidder from continuing in the competition and major or "fatal" deficiencies which cause the proposal to be immediately rejected.
- Adjectival Rating: Adjectival rating systems utilize a specific set of adjectives to describe the conformance of an evaluated area within a proposal to the project's requirements in that area. Adjectival rating systems are an extension of modified satisficing. They recognize that a more descriptive rating system is in order and that the rating system should be continuous rather than discreet.
- Direct Point Scoring: Direct point scoring evaluation allows for more rating levels and thus may appear to give more precise distinctions of merit. However, it may lend an unjustified air of precision to evaluations. Evaluators assign points to evaluation criteria based upon either some predetermined scale or the preference of the evaluator. This system is used by many transportation agencies, but is out of favor with many other federal agencies because the use of numerical rating systems in conjunction with specific percentage weightings for the factors provides the least flexibility the source selection authority in making award decisions.

Table 3. Case Study Project Example Best-Value Rating System Usage.

Best Value Rating System	Example	Remarks/Source
Satisficing	**Responsive:** Meets published criteria. **Non-responsive:** Does not meet published criteria.	Minnesota DOT Highway Project RFP, (MnDOT, 2002)
Modified Satisficing	**Green:** Acceptable – The proposal essentially satisfies the standards; minor weaknesses, even if not corrected, do not render this proposal/factor unacceptable. **Yellow:** Marginal – Reasonably susceptible to becoming acceptable. The proposal/ factor fails to adequately satisfy the standards. However, significant weaknesses/ deficiencies can be corrected through exchanges. Weaknesses/Deficiencies are such that failure to correct may render this major proposal/factor unacceptable. **Red:** Unsatisfactory – The proposal fails to meet stated criteria and is not capable of becoming acceptable without major revisions.	U.S. Army Corps of Engineers Air Freight Terminal/Airfield project RFP (USACE, 2002)
Adjectival Rating	**Excellent** - Of exceptional merit; exemplary performance in a timely, efficient, and economical manner. Performance which, in addition to fully satisfying contract and/or customer requirements, features above-average innovation or efficiency and rare or nonexistent deficiencies. **Very Good** - Very effective performance, which is fully responsive to contract or customer requirements, accomplished in a timely, efficient and economical manner; for the most part; only minor deficiencies; deficiencies do not affect overall performance. **Good** - Effective performance, fully responsive to contract requirements, reportable deficiencies, but with little identifiable effect on overall performance. **Satisfactory** - Meets or slightly exceeds minimum contract requirements, reportable deficiencies, but with little identifiable effect on overall performance. **Poor/Unsatisfactory** - Performance does not meet minimum acceptable standards, fails to meet contract requirements and/or customer expectations and which includes deficiencies that impact other areas of work performance.	NASA Johnson Space Center Tunnel System Project RFP (NASA, 2001)
Direct Point Scoring	**Marginal** - 0-70 points **Satisfactory** – 70 – 90 Points **Excellent**- 90 – 100 points	Maine DOT Bridge Project RFP (MDOT, 1994).

Figure 2 illustrates the continuum formed by the available best-value rating systems. It shows that as project complexity increases, that the complexity of the best-value rating system can increase to accommodate multiple competing objectives within a given project. Thus, it can be seen that each project will have a rating system that is best suited for its particular needs and that an agency's attempt to standardize rating systems for all projects may not be a wise decision.

Satisficing	Modified Satisficing	Adjectival Rating	Direct Point Scoring

◄	►
Simple	Complex
Quick	Requires Analysis
Bimodal Outcome	Array of Outcomes
Assessment Accuracy not Critical	Assessment Accuracy Critical

Figure 2. Best-Value Rating System Continuum.

Best-Value Award Algorithms

Best-value award algorithms define the steps that owners take to combine the parameters, evaluation criteria, and evaluation rating systems into a final award recommendation. Seven best-value award algorithms have been discovered through a comprehensive analysis of the literature and project procurement documents. A description of each of these procedures follows. The algorithms are described through the formulas illustrated in Table 4. The seven algorithms are:

- Meets Technical Criteria–Low Bid: The final award decision is based on price. Technical proposals are scored before any cost proposals are reviewed. The price proposal is opened only if technical proposal is above the minimum technical score. If it is below the technical score, the proposal is deemed non-responsive and the price proposal is not considered. Award will be determined by the lowest priced, fully qualified offeror.
- Adjusted Bid: Price proposals are opened after the technical proposals are scored. When the price proposal is opened, the project price is adjusted in some manner by the technical score, typically through the division of price by a technical score between 0 – 1. The adjusted bid is used only for project award. The offeror will be paid according to the price stated in the price proposal. The offeror with the lowest adjusted bid will be awarded the project.
- Adjusted Score: The price proposals are opened after the technical proposals are scored. The adjusted score is calculated by multiplying the technical score by the total estimated project price and then divided by the price proposal. The award is to the offeror with the highest adjusted score.

Table 4. Best-Value Award Algorithm Details and Case Study Project Usage.

Best Value Award Algorithm	**Algorithm**	**Variables**	**Case Study Project Use**
Meets Technical Criteria–Low Bid	If $T > T_{min}$, **Award to P_{min}** If $T < T_{min}$, Non-Responsive	T = Technical Score; P = Project Price	8
Adjusted Bid	AB = P/T **Award AB_{min}**	AB = Adjusted Bid; T = Technical Score; P = Project Price	7
Adjusted Score	AS = (T x EE)/P **Award AS_{max}**	AS = Adjusted Score; EE = Engineer's Estimate; T = Technical Score; P = Project Price	7
Weighted Criteria	$TS = W_1S_1 + W_2S_2 + \ldots + W_iS_i + W_{(i+1)}PS$ **Award TS_{max}**	TS = Total Score; W_i = Weight of Factor $_{i,}$ S_i = Score of Factor $_{i;}$ PS = Price Score	12
Quantitative Cost-Technical Trade-off	$T_{Increment} = [(T_j/T_j) - 1] \times 100\%$ $P_{Increment} = [(P_j/P_i) - 1] \times 100\%$ **If $T_{Increment} > P_{Increment}$, Award Proposal$_i$** If $T_{Increment} < P_{Increment}$, Do Not Award Proposal$_j$, Repeat with Proposal$_{j+1}$ **Repeat Process until $T_{Increment} > P_{Increment}$**	T = Technical Score; P = Project Price;	4
Qualitative Cost-Technical Trade-off	Similar to above, only no quantitative analysis of difference. **Award to proposal that has best value in proposed scope.**		11
Fixed Price–Best Proposal	**Award T_{max}, Fixed P**	T = Technical Score; P = Project Price;	1

- Weighted Criteria: The technical proposal and the price proposal are evaluated individually. A weight is assigned to the price and each of the technical evaluation factors. The sum of these values becomes the total score. The offeror with the highest total score is selected.

- Quantitative Cost-Technical Trade-off: This involves calculating the technical score and the price score increment and then examining the difference between the incremental advantages of each. The increment in the technical score is calculated by dividing highest technical score by the next highest technical score less one multiplied by 100%. The increment in price score is calculated dividing highest price score by the next highest price score less one multiplied by 100%. The award is made to the offeror with the lowest price, unless the higher priced offers can be justified through a higher technical value. This justification is made by determining if the added increment of price is offset by an added increment in technical score.

- Qualitative Cost-Technical Trade-off: This method relies primarily on the judgment of the selection official and not on the evaluation ratings and scores. The final decision consists of an evaluation, comparative analysis, and trade-off process that often require subjectivity and judgment on the part of the selecting official.
- Fixed Price–Best Proposal: This method utilizes a maximum price or a fixed price for the project. Offerors must submit price proposal that is equal to or less than the specified bid price. The award is based only on the technical proposal evaluation. The offeror selected will be that whose technical score is the highest.

The qualitative cost-technical tradeoff and the weighted criteria algorithms are the most frequently used and make up nearly one-half (23 of 50) of the sample population. The adjusted score, adjusted bid, and meets technical criteria–low bid algorithms are approximately equal in number and comprise 44% of the sample. The quantitative cost-technical tradeoff and the fixed-price–best proposal algorithms are used only 4% of the sample.

The best-value parameters, evaluation criteria, evaluation rating systems and award algorithms described in this section are a generic synthesis of what the entire design and construction industry defines as best-value procurement. The differences in concepts are found due to the agencies that use them and some are due to the nature of the projects themselves. As a next step, the research team benchmarked the current practices in the highway industry against those in the general construction industry. The next section discusses the results of a survey regarding the use of best-value in the highway construction industry.

National Transportation Agency Survey Results

As outlined in the methodology, the research team developed a survey to obtain information related to the state of practice of best-value procurement in the highway construction industry. Of the 41 agency representatives responding, 27 respondents answered that the agency had some experience with best-value procurement, two (2) agency representatives responded that the agency had no experience but planned to use best-value in the near future, and 12 respondents indicated that the agency had no experience with best-value procurement. The answers to this question revealed that among the respondents, the majority (66%) of agencies had experience with some form of "best-value" procurement.

The second question asked respondents to define the particular selection strategy or strategies used among the methods defined in the questionnaire. The following summarizes the variety of selection strategies used and the frequency of their use:

- 10 of 27 use "Meets Tech. Criteria-Low Bid" (37%)
- 7 of 27 use "A+B" (20%)
- 6 of 27 use "Adjusted Bid" (22%)
- 6 of 27 use "Weighted Criteria" (22%)

- 3 of 27 use "Multi-parameter" (11%)
- 2 of 27 use "Cost/Technical Trade off" (7%)
- 1 of 27 use "Adjusted Score" (5%)

The responses indicated that the best-value selection strategy used most often (37%) was Meets Technical Criteria–Low Bid. Several respondents included A+B Bidding and Multi-Parameter Bidding as selection strategies in the "Other" category. If these strategies are assumed to be equivalent as noted in the definition, the Multi-Parameter strategy was the next most frequently used strategy (31%). This distribution indicates that the best-value selection strategies adopted by transportation sector agencies are more closely aligned with the low bid system compared to the distribution of the award methods of the larger sample of projects, including vertical projects and projects outside of the transportation sector presented in previous sections of this chapter. The larger sample population of case study project RFPs presented in Table 4 indicated that the weighted criteria and cost-technical tradeoff strategies were the most frequently used, comprising half (25 of 50) of the sample population.

The third question asked respondents to identify what key criteria were used by the agency in the qualification or selection process. The key criteria and frequency of their use are summarized as follows:

- 16 of 25 use "Past Performance" (64%)
- 15 of 25 use "Projected Time" (60%)
- 13 of 25 use "Personnel Qualifications" (52%)
- 11 of 25 use "Management Capabilities" (44%)
- 6 of 25 use "Public Interface Plan" (24%)
- 6 of 25 use "Technical Capability/Solutions" (24%)
- 9 of 25 use other categories (36%)

The survey results for the transportation agencies indicate that past performance and projected time are the most frequently used criteria followed by qualifications of personnel. In comparison, the larger sample population cited past performance and qualifications of key personnel as the most frequently used criteria. In the case of transportation agencies, it appears that projected time performance is more important than other commonly used criteria.

Summary and Conclusions

This paper has defined the state of the industry for best-value procurement methods. Current trends in legislation are paving the way for wide-spread use of best-value procurement for highway construction projects. Four key best-value concepts of parameters, evaluation criteria, evaluation rating systems and award algorithms have been defined in this research and presented in this chapter. The application of these concepts was validated through 50 summary level and 14 detailed best-value case studies from all sectors of public construction. Lastly, best-value use in the highway industry was benchmarked though a nation-wide survey of state transportation agencies. This details the state of the industry and has provided the framework for a critical analysis of best-value methods for potential use in highway construction projects.

Several conclusions can be drawn from the results of the above analyses. First, the best-value parameter-based framework developed to describe the fundamental elements of this type of contracting works well to distill the essence of a given project and describe the salient contractual mechanisms that are inherent to its make-up. This can be extended from the analysis of case studies to the development of new best-value project procurement documents. Public owners should first identify those parameters that are of specific interest in a given project. Next, for each of the parameters, best-value evaluation criteria should be generated to permit the evaluation of competing proposals. To do this, a best-value rating system and best-value award algorithm must be selected from among the options for each identified in this paper. All of the above can then be published in the best-value RFP making the method by which the best-value proposal is determined completely transparent to the competitors. By following this approach, public owners will ensure that the entire best-value process is covered and that justification for each component in the ultimate selection decision can be justified before the RFP is published.

Next, it can be concluded that best-value contracting is already is being used successfully in the industry. The number of best-value procurement documents that were found combined with the excellent response rate to the survey both indicate that public owners not only understand the potential advantages of using this method but are able to convert these theories into practice in their construction programs.

Finally, the intersection of the literature, the case study project contents, and the survey results seem to indicate that the most important factor to public owners is the qualifications and past performance of the construction contractor. This is followed by contract price and schedule. Thus, it can be concluded that the major impetus behind implementing best-value contracting is to ensure that a competent contractor with a track record of success wins the project rather than merely the one who may have made the biggest mistake of its cost estimate. Owners apparently are willing to pay a marginal amount more for their construction to achieve this objective with regard to the "quality" of the contractor that will build their project.

Acknowledgements

The authors would like to gratefully acknowledge their appreciation to the NCHRP, Transportation Research Board, under the National Academy of Sciences for sponsoring this research. Publication of this paper does not necessarily indicate acceptance by the Academy of its contents, either inferred or specially expressed herein. The authors would like to thank the NCHRP Project 10-61 Technical Panel for their comments and direction during the research process, as well as all of the respondents who contributed survey data to this project. Finally the authors would like to thank all of the research team's advisory panel and the student research assistants who worked so diligently on this project.

References

Army Source Selection Guide. (2001). Report, U.S. Army Corps of Engineers, Office of the Assistant Secretary of the Army, Acquisitions, Logistics and Technology. http://www.amc.army.mil/amc/rda/rda-ap/ssrc/fr_ssll.htm, January 15, 2003.

Colorado Revised Statutes (2002). Title 24, Article 103 Source Selection and Contract Formation, Part 2 "Methods of Source Selection," 24-103-202.3.

Delaware Code (2001). Title 29, Part VI Budget, Fiscal, Procurement And Contracting Regulations Chapter 69 State Procurement Subchapter IV. "Public Works Contracting," [As Amended By 73 Delaware Laws 41 (2001)]. 29 Del. C. § 6962.

Dorsey, R. (1995) "New Paradigms in Construction," presented at Associated General Contractors of America, Project Delivery Systems for Building Construction Conference, Detroit, Michigan, October 26-27, 1995.

Federal Acquisition Regulation (FAR). (2000). "Part 15 - Contracting By Negotiation," U.S. Government Printing Office, Washington, D.C.

Federal Highway Administration (FHWA) (1998). "FHWA initiatives to Encourage Quality Through Innovative Contracting Practices Special Experimental Projects No.14 – (SEP-14)", U.S. Department of Transportation, http://www.fhwa.gov///programadmin/contracts/sep_a.html (October 23, 1998)

Kentucky Revised Statutes (2002). Kentucky Model Procurement Code, Chapter 45A, "Competitive Sealed Bidding." 45A.080.

Maine Department of Transportation (MDOT) (1994). "Request for Proposals: Bath/Woolwich Design-Build Bridge Project," Augusta, Maine.

Minnesota Department of Transportation, (MnDOT) (2002). "T.H. 100 Design-Build Request For Proposals," State Project 5502-85, Minnesota Department of Transportation, St. Paul, Minnesota

National Aeronautics and Space Association (NASA) (2001). "Request for Proposals: Johnson Space Center, Tunnel System Design-Build Project," RFO # 9-BJ33-T13-0-03P, Houston Texas.

National Cooperative Highway Research Program (2001). Guidelines for Warranty, Multi-Parameter, and Best Value Contracting, NCHRP Report 451, Transportation Research Board, National Research Council, Washington, D.C.

U.S. Army Corps of Engineers, (USACE) (2002). "Request for Proposals: Air Freight Terminal/Airfield Project," New York, New York.

U.S. Postal Service (2000). Handbook, Design and Construction Purchasing Practices U.S. Government Printing Office, Washington, D.C.

U.S. Department of Justice, (2000). Federal Bureau of Prisons, Program Statement: Design & Construction Procedures, U.S. Government Printing Office, Washington, D.C.

Appendix 1: Case Study Summary

Table 1.1 illustrates the additional information gleaned from the analysis of best-value RFP's collected during the first phase of this study. The case study summary below is based upon the same 50 cases previously presented in Table 2 in the best-value parameter section.

Table 1.1. Best-Value Award Algorithm Case Study Summary

State/Agency	Agency Terminology	Remarks	Best-Value Award Algorithm
Alaska DOT	Criterion Score	Divide Technical Score by Price	Adjusted Score
Arizona DOT	Quality Adjusted Price Ranking	Percentage system used to adjust bid price for technical score	Adjusted Bid
Colorado DOT Pre-1999	Low Bid, Time Adjusted	Multi-parameter bid with qualifications	Meets Technical Criteria – Low Bid
Colorado DOT Post-1999	Best-value	May use weighted criteria to arrive at an adjusted score	Adjusted Score
Delaware DOT	Competitive Proposals	Design Alternates, Qualifications, Scheduled, and Price scored.	Weighted Criteria
District of Columbia DPW	Best-value	Adds owner contract administration costs to price	Adjusted Score
Florida DOT	Adjusted Score	May also include time adjustment	Adjusted Score
Georgia DOT	Low Bid, Prequalified	Short list by qualifications	Meets Technical Criteria – Low Bid
Idaho DOT	Weighted Selection	Cost 51%; Qualifications/ Past Experience 49%	Weighted Criteria
Indiana DOT	Low Bid, Fully Qualified	Minimum technical score to be found qualified	Meets Technical Criteria – Low Bid
Maine DOT	Overall Value Rating	Divide Price by Technical Score	Adjusted Bid
Massachusetts DOT	Best-value	Included life-cycle cost criteria	Weighted Criteria
Michigan DOT	Low Composite Score	Divide Price by Technical Score	Adjusted Bid
Minnesota DOT	Low Bid, Fully Qualified	Short list by qualifications	Meets Technical Criteria – Low Bid
Missouri DOT	Low Bid + Additional Cost	Additional costs include life-cycle cost calculation	Meets Technical Criteria – Low Bid
New Jersey DOT	Modified Low Bid	Included design costs	Meets Technical Criteria – Low Bid
North Carolina DOT	Quality Adjusted Price Ranking	Percentage system used to adjust bid price for technical score	Adjusted Bid
Ohio DOT	Low Bid	Includes design costs	Meets Technical Criteria – Low Bid
Oregon DOT	Best-Value	Combine technical with cost by weights	Weighted Criteria
South Carolina DOT	Low Composite Score	Divide Price by Technical Score	Adjusted Bid
South Dakota DOT	Best-Value	Divide Price by Technical Score	Adjusted Bid

State/Agency	Agency Terminology	Remarks	Best-Value Award Algorithm
Utah DOT	Best-Value	Combine technical with cost by weights	Weighted Criteria
Virginia DOT	Two Step Selection	Qualifications/Experience in Step 1 and Price and Technical in Step 2	Weighted Criteria
Washington DOT	High Best-Value Score	Divide Technical Score by Price	Adjusted Score
Alberta, Canada, Ministry of Highways	Value Index	Divide Technical Score by Price	Adjusted Score
City of Reno, Nevada	Best-Value	Qualifications & Past Performance equal to Price	Weighted Criteria
City of Santa Monica, California	RFP Process	Requires Guaranteed Maximum Price and life-cycle criteria	Qualitative Cost-Technical Trade-off
City of Wheat Ridge, Colorado	RFP Process	Uses Weighted Criteria approach to arrive at technical score	Fixed Price/Best Design
District of Columbia Schools	Best-Value	Responsiveness check for qualifications, experience & subcontracting plan. Award to lowest, fully responsive bid.	Meets Technical Criteria – Low Bid
Federal Bureau of Prisons	Best-Value	Uses Weighted Criteria approach to arrive at technical score	Qualitative Cost-Technical Trade-off
Federal Highway Administration	Best-Value	Adds owner contract administration costs to price. Uses Adjusted Score formula to differentiate between bids	Quantitative Cost-Technical Trade-off
Fort Lauderdale County, Florida	Selection/Negotiation	Requires Guaranteed Maximum Price	Weighted Criteria
General Services Administration	Best-Value	Uses Weighted Criteria approach to arrive at technical score	Qualitative Cost-Technical Trade-off
Los Alamos National Laboratory	Best-Value	Two phase selection	Weighted Criteria
Maricopa County, Arizona	Quality Adjusted Price Ranking	Uses Weighted Criteria approach to arrive at technical score. Then computes a "$-value" of technical proposal and subtracts from price	Adjusted Bid
Naval Facilities Engineering Command	Best-Value	Uses Weighted Criteria approach to arrive at technical score	Qualitative Cost-Technical Trade-off
Nashville County, Tennessee	Competitive Sealed Proposals	Qualifications, Management Plan and Price plus Warranty	Adjusted Score
National Aeronautics and Space Administration	Best-Value	Uses Weighted Criteria approach to arrive at technical score	Qualitative Cost-Technical Trade-off

State/Agency	Agency Terminology	Remarks	Best-Value Award Algorithm
National Institute of Standards and Technology	Best-Value	Uses Weighted Criteria approach to arrive at technical score.	Qualitative Cost-Technical Trade-off
National Park Service	Best-Value	Uses "technically acceptable" approach to arrive at technical score.	Qualitative Cost-Technical Trade-off
Pentagon Renovation Program Office	Best-Value	Uses Weighted Criteria approach to arrive at technical score; includes incentive clauses.	Qualitative Cost-Technical Trade-off
Seattle Water Department	Best-Value	Uses Weighted Criteria approach to arrive at technical score	Quantitative Cost-Technical Trade-off
University of Colorado	Best-Value	Qualifications/Experience in Step 1 and Price and Technical in Step 2	Weighted Criteria
University of Nebraska	Best-Value	Qualifications/Experience in Step 1 and Price and Technical in Step 2	Weighted Criteria
US Army Corps of Engineers	Best-Value	Uses Weighted Criteria approach to arrive at technical score	Qualitative Cost-Technical Trade-off
US Customs Service	Best-Value	Uses Weighted Criteria approach to arrive at technical score. Requires Guaranteed Maximum Price	Qualitative Cost-Technical Trade-off
US Department of Energy	Best-Value	Uses Weighted Criteria approach to arrive at technical score	Quantitative Cost-Technical Trade-off
US Forest Service	Best-Value	Uses Adjusted Bid formula to differentiate between bids	Quantitative Cost-Technical Trade-off
US Postal Service	Best-Value	Uses Weighted Criteria approach to arrive at technical score	Qualitative Cost-Technical Trade-off
Utah Dept. of Natural Resources	Value Based Selection	Combine technical with cost by weights	Weighted Criteria

Preference for A + B Contracting Technique among State Departments of Transportation

Kelly C. Strong,[1] Nolan Raadt,[2] and James Tometich [3]

Abstract

The objectives of the research are to compare performance, cost and value implications of design-build contracts, A+B contracts, lane rental contracts and traditional contracts. Specific performance and cost measures considered are Administration Costs, Project Costs, Management Complexity, Disruption to Third Parties, Road User Costs, Innovation, Product/Process Quality, and Funding Flexibility & Duration. Performance parameters are compared on nine different project types. The research compares the three innovative contracting techniques to traditional contracting on relevant performance factors for each project types, resulting in a "best practices guide" along with project selection criteria for innovative contracting methods. The research methodology utilized a survey of national experts who rated each innovative contracting method for each performance factor on each of the project types. Results of the findings from the survey of national experts as well as summaries of case study interviews are described and discussed.

Keywords- A + B Contracts; innovative contracting; performance

Problem Statement

Many governmental agencies charged with delivering public infrastructure are experimenting with innovative contracting methods and have been over the past ten years. Many of the more common techniques have recently been formally approved for use by the Federal Highway Administration (FHWA 2002). One particular federal program, Special Experiment Project Number 14 (SEP-14), is helping to accurately define and clarify many of these new innovative contracting methods to ensure that the processes and practices involved with innovative contracting are implemented effectively. The mandates of SEP-14 apply only to federally funded projects, but states can use the techniques presented in the SEP-14 legislation on state funded projects. The primary objective of SEP-14 is to review specific innovative contract techniques as they are applied to specific projects, which are monitored closely by participating state Departments of Transportation (DOT) to measure the

---

[1] Assistant Professor , Department of Civil, Construction and Environmental Engineering, Iowa State University, 454 Town Engineering, Ames IA 50011, 515-294-1460, Fax: 515-294-3845, kstrong@iastate.edu

[2] Research Assistant, Department of Civil, Construction and Environmental Engineering, Iowa State University, 454 Town Engineering, Ames IA 50011, 515-294-1460

[3] Research Assistant, Department of Civil, Construction and Environmental Engineering, Iowa State University, 454 Town Engineering, Ames IA 50011, 515-294-1460

effectiveness of innovative contracting compared to the traditional design-bid-build method or other acceptable methods. However, many states are lagging in report preparation, and those states that have completed reports have not presented their information in a consistent manner.

The specific innovative contracting methods under investigation in the SEP-14 report are:

- A + B with an Incentive/Disincentive option (A + B w/ I/D),
- Lane Rental,
- Warranty Clauses, and
- Design-Build.

A + B contracting is sometimes referred to as cost plus time contracting, or biparameter bidding. In A + B contracting, the submitted bids include a cost and a schedule for completing the work, typically within some boundary conditions established by the DOT. The DOT can then translate time differences into an economic value by using the average daily Road User Costs in order to optimize the economic value (costs/benefits) of the bids. The DOT may also include incentive and/or disincentive (I/D) language in the contract as motivation to complete the project in a timely manner. Incentive clauses indicate the bonus to be paid if the contractor finishes earlier than the contractual completion state, whereas disincentive clauses indicate penalties to be imposed for failing to meet the contract completion date. The award is made to the bidder whose combination of cost and time reflects the best value. Road User Costs and the method for incorporating them into the award decision are discussed in more detail in subsequent sections of the paper.

Lane rental is somewhat similar to A + B contracting in that attempts are made to include the "cost" of disruptions or loss of service to the traveler in the bid consideration. Bidders include a schedule for lane or shoulder closings along with a cost for construction. The DOT assigns an hourly or daily charge for closing a lane or shoulder and charges the contractor for the closure. The objective is to motivate contractors to minimize closures on high-volume roadways or during high volume times. Similarly to A + B contracts, Road User Costs form the basis of the lane rental fees.

Warranty clause projects utilize contracts that shift the burden of maintenance and repair of deficiencies to the contractor for a specified period after completion of the project. A warranty contract transfers much of the quality process responsibility to the contractor, along with the risk of quality failures. Because warranty clause contracts are very dissimilar to the other innovative contract types, they were excluded from consideration in this study.

Design-build is a contract method where a single entity is awarded a contract for both design and construction of the project. Whereas A + B and lane rental contracting fit the traditional delivery model of design-bid-build, design-build represents an alternative form of delivery where construction begins prior to the completion of design. Therefore, design-build is a departure from both traditional procurement and delivery. Although design-build can involve complex funding, procurement and delivery processes, there are many benefits to DOT's from the use of design-build. The most significant of these benefits is the potentially drastic reduction in overall project time.

After a five-year evaluation period, the FHWA declared on May 4, 1995 that A + B contracting (both with and without I/D) and lane rental contracts are operational and should no longer be considered experimental. On December 10, 2002, the FHWA published the final rule for design-build, making it operational for use in federally-aided highway projects.

Since innovative contracting methods have been practiced for several years in many states, and the federal government has recognized and defined many standard practices for innovative contracting, the industry can benefit from an examination of the preferences for using innovative contract options among state DOT's. Several states have researched innovative contracting methods with the objective of developing a protocol to assist agency personnel in selecting the most effective contract type based on certain project parameters. There also have been reports by various non-governmental organizations and institutions that have researched one or more innovative contracting technique. The main reports and most comprehensive studies are outlined in the following paragraphs in order to develop an integrated summary and synthesis of current thinking on the comparative effectiveness of innovative contracting methods.

The American Association of State Highway and Transportation Officials (AASHTO) 2001 Primer on Contracting for the 21st Century reports that the standard practice for A + B contracting is to use calendar days for the "B" term to avoid possible confusion and controversy over duration and to use the formula A+ (B * RUC/Day) as the basis for award consideration, where A represents cost in dollars, B represents contract days, and RUC/day represents daily Road User Costs. Road User Cost is an estimated cost to the traveling public that arises from detours and less efficient traffic routing that add travel time as well as the costs associated with diminished road capacity and the resultant slow traffic. The AASHTO Primer also encourages the use of I/D clauses to keep the playing field level and prevent artificially aggressive "B" term offers by bidders. AASHTO suggests that A + B contracts can be an effective practice for critical projects that have high road use delay impacts.

The AASHTO Primer reports that an informal survey in 1998 by the FHWA indicates that 18 states are using A + B contracts. By 2001, 28 states plus the District of Columbia were using A + B projects as part of SEP-14 reviews, and another 12 states were using A + B outside of SEP-14 participation. Maryland, Missouri, Florida, and New York have been most active in use of A + B contracts. Recent surveys suggest that virtually all states have adopted some degree of time-based contracting. A + B contracting is frequently couple with I/D clauses, although A + B awards can be made without I/D clauses, and I/D clauses can be used with any contract type. I/D amounts can be based on estimates of such items as traffic safety risks, maintenance of traffic costs, or road user delay costs. Florida has experimented with variable I/D amounts relative to the order of magnitude of the delay or acceleration. With variable I/D, the amount is increased geometrically further from the target completion date.

The Michigan DOT reported in February 2000 that I/D clauses had been utilized on 26 projects (AASHTO 2001). Of these 26 projects, 65% were completed early, 12% were completed on time, and 23% were completed late. The average net reduction in schedule duration days was 19% compared to similar projects without I/D clauses. The average increase in expenditures as a result of incentive payments was 1.5% of contract amount.

In one of the earliest reviews of innovative contracting, the South Dakota DOT hired Trauner Consulting Services, Inc. to assist them in defining criteria and guidelines that South Dakota could use to determine the most effective innovative contracting methods (South Dakota DOT 1996). The South Dakota report indicates A + B contracting is likely to result in higher construction costs, especially if I/D clauses are incorporated into the contract. Additionally, A + B contracting will likely result in higher internal costs to the DOT, but construction time will be shortened. Case reviews suggest that the quality of workmanship remained unchanged with some reports of improved quality. The South Dakota report claims that A + B is the lowest risk method of the time-based contracts studied (I/D, lane rental, and A + B) because time is part of a competitive bid review and schedule risk is passed on to the contractor in the form of determining optimal combinations of price and schedule. The Trauner report to the South Dakota DOT makes specific recommendations about the use of A + B contracts, recommending use for projects with high traffic volumes or restrictions on interstates or principal arterials with Road User Cost estimates of at least $2500 more than the liquidated damages on the project. They also recommend A + B for projects with long detours causing delays in excess of 10 minutes for the traveling public or for projects where high accident rates or worker safety concerns are prominent. A third type of project appropriate for A + B contracting is one with significant impact to the local community or economy. Finally, any project that must be coordinated with special events is a good candidate for A + B contracting. Trauner suggests that South Dakota DOT have all Right-of-Way (ROW) acquisition, related contracts, utility issues, and other potential conflicts resolved prior to issuance of the bid documents. A + B can also be helpful when DOT personnel seek contractor input on phasing of work and optimal scheduling (South Dakota DOT 1996).

The Ohio DOT has internally developed a manual (Ohio DOT 2003) to assist in developing construction contracts through the help of a selection criteria process based on project parameters. A + B is listed as appropriate for all types of projects, although each project must be evaluated independently in order to choose the best contracting method based on project specifics.

The Utah Technology Transfer Center generated a best practice guide for innovative contracts (Bolling & Holland 2003) suggesting that A + B contracting is most appropriate for projects with traffic restrictions, lane closures, and long detours resulting in Road User Costs. The Utah Technology Transfer Center report also notes that conflicts from utility relocations, design uncertainty, environmental factors, or ROW acquisition can drastically reduce the effectiveness of A + B contracting. The report further suggests that the only instance that I/D should NOT be included is when specific completion dates are flexible and when Road User Costs are not severe.

A study by the National Cooperative Highway Research Program (NCHRP 2001) states that shortened schedules are the primary benefits of A + B contracts, especially if incentive clauses are included. Use of A + B contracts also facilitates innovation by contractors for more efficient project delivery. Disadvantages of A + B reported in the NCHRP study include greater burden on state agency resources due to extended work hours and additional personnel on site for inspections and testing.

The Florida DOT (Florida DOT 2000) reported that cost and time overruns are reduced using A + B contracts, averaging 3.5 % cost overrun and 8.1% time overrun compared to 12.4% cost overrun and 30.7% time overrun on traditional low-bid award projects. These data are from nine A + B projects during the 1997-1998 period.

The Battelle group (Carpenter, Fekpe, and Gopalakrishna 2003) reported that A + B is typically used by state agencies for reconstruction, rehabilitation, and remediation projects in urban settings where the public impact is high. Primary motivations for use of A + B and other innovative contracting methods include reduced public impact, minimization of traffic disruption, ability to shift risk and liability to contractors, and improved overall product quality and lower total costs (including Road User Costs and social impacts). The Battelle report stated that A + B was the most common innovative contracting method used by the states (along with short term warranties).

Shen, Drew and Zhang (Shen, Drew, & Zhang 1999) argued that the value of time to the client, typically expressed as daily Road User Costs, must be compared to the value of time to the contractor, expressed as the price/time performance curve derived from the A and B terms of the bid, in order to choose optimal bid parameters. In other words, the cost per unit of time to society and the cost per unit of time to road users must be balanced to optimize resources

A study by Shr, Thompson, Russell, Ran, and Tseng (Shr, et al 2000) examined A + B contracting and determined that some loss of value or suboptimum contracting is possible if state DOT's do not place an upper and lower limit on the time parameter of the bid. Shr, Ran, and Sung followed up this study in 2004 (,Shr, et al 2004) adding that the factors of I/D costs should be added to Road User Costs and then optimized against the A + B (cost plus time) parameters in each of the bids received in order to choose the lowest total cost option. This optimization process shifts much of the time risk experienced on a project to the contractor while maximizing the agencies resources. However, the optimization modeling appeared to be cumbersome for use on a regular basis across a variety of projects.

In summary, the literature suggests that A + B contracting is effective in reducing project schedules, is most appropriate for projects with high Road User Costs or traffic restrictions, that sources of conflict (utilities, ROW, environmental issues, etc.) should be resolved prior to the start of construction, and that A + B contracts can result in higher internal administrative costs.

Many of the studies referenced above use either project characteristics (high Road User Costs, project complexity, social impacts) or performance criteria (cost, schedule, I/D), but few use both. Also, most of the studies were single agency studies. Lastly, some of the conclusions were derived from logical analysis or anecdotal evidence, with few comprehensive confirmations of assumptions. The research project described in this paper attempts to create a balanced analysis of the preferences contracting methods by asking national experts to rank order four contracting options utilizing both performance and project characteristics. After comparison of preferences for different innovative contracting methods using national survey data, the research study examined case study information for evidence of increased administrative costs, construction costs, and construction time, along with other potential issues that could affect the decision to use A + B contracting.

Research Objective

The purpose of this research project is to examine the preferences of state departments of transportation for using different forms of innovative contracting compared methods and to provide transportation managers and educators with insight and recommendations for the use of innovative contracting in transportation policy. The research project is comprised of two principal components, a national survey of state DOT construction engineers and a more in-depth case study of three A + B projects. The purpose of the national survey is to provide insight into the preferences of state DOT construction engineers regarding the project and performance factors that appear to favor different innovative contracting forms. Because the findings from the national survey showed such a strong preference for A + B contracting, the research team conducted a second component of the research effort involving in-depth case study of three A + B jobs conducted in the state of Minnesota in order to investigate the project nuances that may not be discovered in the national survey.

Research Methodology

The research methodology in this study utilized multiple methods of analysis incorporating both qualitative and quantitative techniques. The first step in the methodology was to identify relevant performance criteria. The Minnesota DOT (Mn/DOT) identified relevant performance factors to be used in determining project success (Minnesota DOT 2003). The research team chose a subset of these performance factors relating to construction procurement and contracting value. The eight relevant performance factors are:

- A*dministrative Costs,* defined as the different types of internal costs Mn/DOT incurred in tracking processes: contract administration, inspections, reviews, ROW acquisition, and environmental assessment and monitoring.
- *Construction Costs,* including first costs, costs of change orders, cost of engineering and design, and environmental remediation.
- *Time,* referring to the overall length of time spent in project-specific planning, funding/appropriations, design, construction, and extensions.
- *Management Complexity,* referring to the relative difficulty of coordinating issues encountered over the course of the project, specifically management-related aspect of the project such planning and establishment of scope,

logistical challenges, utility relocation and coordination, adjustments to unforeseen problems that arose during execution of the project, etc.

- *Third Party Impacts,* including disruptions to businesses, schools, churches, residential neighborhoods, and other establishments or destinations along the route.
- *Road User Costs,* referring to the costs incurred by the motoring public resulting from the project. Some examples of Road User Costs include accidents, driver time, and additional vehicle mileage resulting from detours.
- *Quality,* referring to the level of workmanship and the end products' performance versus what is expected by the owner, as well as the amount of post-construction call-backs and required maintenance of the facility.
- *Innovation,* including the degree to which the contractor is able to use new or less conventional concepts, methods, or materials on the project, the contractor's flexibility to make design changes and pursue alternative ideas or techniques aimed at reducing cost and schedule.

The researchers were interested in determining which types of projects would be candidates for use of different innovative contracting forms in the opinion of state DOT construction engineers. To determine this, state DOT construction engineers are asked to compare traditional design-bid-build, A + B contracting, lane rental, and design-build by ranking the four contracting methods for each performance factor on a variety of project types. The participating states are:

Arizona	Arkansas	Colorado	Florida
Georgia	Idaho	Maryland	Michigan
Mississippi	Nebraska	New York	Oregon
South Carolina	South Dakota	Tennessee	Texas
Virginia	Wisconsin	Wyoming	

All of the participating states have used innovative contracting. Five of the nineteen states have used A + B, lane rental, and design-build, nine states have used two of the three innovative contract types, and five states have used one form of innovative contracting.

The research team wanted a cross-section of project types involving a range of anticipated Average Daily Traffic (ADT) counts, construction complexity, management-of-traffic issues, and design efforts. The project examples include:

- Major corridor realignment/expansion
- Multi-lane highway rehabilitation through a city, with detours
- Multi-lane highway rehabilitation through a city, under traffic
- Rural bridge replacement
- Metropolitan bridge replacement
- 2-lane highway resurfacing
- Mill and overlay
- Unbonded concrete overlay

- Preservation project with culvert replacement during 2-lane highway resurfacing

Not all performance criteria are of equal importance on the different projects types. Therefore, a minimum 50% subsample (9 of 19) of the responding DOT construction engineers was surveyed to determine the relative importance of each performance factor for each of the project types. Because weighting of performance criteria is more time consuming than the rating of contract types, the research team was concerned that asking the total sample to provide weights would result in a reduction in overall response rates, and increase in cost, and a lengthening of the data collection period. Subsampling is becoming a common practice in survey methodology (Abi-Habib, et al 2002) to reduce costs, increase response rates in the general sample, and shorten data collection times. The DOT construction engineers were asked to distribute 100 points across the 8 performance factors for each of the nine project types. The assignments were aggregated and averaged to generate a weighting coefficient for each performance factor on each of the project types.

The DOT construction engineers were sent blank templates for each project type and asked to rank the four different procurement methods from 1 (best) to 4 (worst) on each performance factor. Nineteen responses were received. The individual rankings were reverse scored (1=4, 2=3, 3=2, 4=1) so that high effectiveness scores would correspond to effective contracting methods. The reverse scored cell matrices were multiplied by the weighting coefficients and summed and averaged across all respondents to create an aggregate mean effectiveness score. The mean effectiveness scores for each contract type were analyzed using a pairwise t-test for comparison of means. The mean effective scores of the four contract methods were listed from highest to lowest, with the highest score reflecting the preferred method, or the method that maximized the performance factors for each project. The rankings are listed in the following section, with statistical significance noted.

In addition to the performance rankings, each respondent is asked to list project criteria favorable to the use of different contracting forms as well as any other issues they believed were critical to effective use of innovative contracting. Because the focus of this manuscript is on the findings regarding preferences for A + B contracting, the summary of responses to the open ended questions regarding A + B contracting from the national DOT sample is listed in the following section.

Lastly, three A + B projects in Minnesota were examined to gain further insight into some of the issues involved in effective use of A + B contracts. The case studies of these projects incorporated interviews with Mn/DOT project engineers along with project cost analysis and comparison to traditional contracting on projects of similar scope. The results of the cost comparisons along with findings from the case study interviews are also presented in the following section.

Key Findings

National Survey Ranking Data. The national survey of state DOT construction engineers asked them to rank four contracting methods (A + B, lane rental, design-build, and design-bid-build) on eight performance criteria for nine different project types. The rank scores were adjusted according to the importance weights for each of the performance criteria, which differed for each project. In other words, the relative importance of individual performance criteria varied based on project type. The aggregate preference rankings are described in the following paragraphs, with statistical significance of mean differences reported in Table 1a-1i.

For *major corridor realignment/expansion projects*, the weighted performance preference rankings are as follows:

1st preference..... A + B Contracting
2nd preference.... Design-Build
3rd preference.....Traditional
4th preference.....Lane Rental

TABLE 1a	A + B
Design-Build	ns
Traditional	$p<0.10$
Lane Rental	$p<0.10$

Table 1a shows the mean differences in weighted performance evaluation scores between A + B and traditional contracting and lane rental contracting were significant at the 0.10 level or lower. The mean difference between A + B and design-build was not significant at the 0.10 level.

For *multi-lane highway rehabilitation projects through cities with detours*, the weighted performance preference rankings are as follows:

1st preference....... .A + B Contracting
2nd preference.... ...Design-Build
3rd preference..... ..Lane Rental
4th preference.......Traditional

TABLE 1b	A + B
Design-Build	$p<0.001$
Lane Rental	$p<0.001$
Traditional	$p<0.001$

Table 1b shows the mean differences in weighted performance evaluation scores between A + B contracting and lane rental, design-build, and traditional were all significant at the 0.001 level.

For *multi-lane highway rehabilitation projects through cities under traffic*, the weighted performance preference rankings are as follows:

1st preference...... A + B Contracting
2nd preference.... ..Lane Rental

3rd preference.......Design-Build
4th preference.......Traditional

TABLE 1c	A + B
Lane Rental	Ns
Design-Build	p<0.001
Traditional	p<0.001

Table 1c shows the mean differences in weighted performance evaluation scores between A + B contracting and lane rental, design-build, and traditional were all significant at the 0.001 level.

For *rural bridge replacement projects,* the weighted performance preference rankings are as follows:

1st preference....... A + B Contracting
2nd preference.... ...Design-Build
3rd preference.......Traditional
4th preference.......Lane Rental

TABLE 1d	A + B
Design-Build	p<0.05
Traditional	p<0.05
Lane Rental	p<0.05

Table 1d mean differences in weighted performance evaluation scores between A + B contracting and lane rental, design-build, and traditional were all significant at the 0.05 level or lower.

For *2-lane highway resurfacing and upgrade projects,* the weighted performance preference rankings are as follows:

1st preference.........A + B Contracting
2nd preference.... ...Lane Rental
3rd preference..... ...Traditional
4th preference........Design-Build

TABLE 1e	A + B
Lane Rental	p<0.01
Traditional	P<0.01
Design-Build	P<0.01

Table 1e shows the mean differences in weighted performance evaluation scores between A + B contracting and lane rental, traditional and design-build were all significant at the 0.01 level or lower.

For *metropolitan bridge replacement projects,* the weighted performance preference rankings are as follows:

1st preference..... ...A + B Contracting
2nd preference.... ...Design-Build
3rd preference..... ...Lane Rental
4th preference........Traditional

TABLE 1f	A + B
Design-Build	p<.005
Lane Rental	p<0.005
Traditional	p<0.005

Table 1f shows the mean differences in weighted performance evaluation scores between A + B contracting and lane rental, design-build, and traditional were all significant at the 0.005 level or lower.

For *mill and overlay projects,* the weighted performance preference rankings are as follows:

1st preference..... ...A + B Contracting
2nd preference.... ...Lane Rental
3rd preference..... ...Traditional
4th preference........Design-Build

TABLE 1g	A + B
Lane Rental	p<0.001
Traditional	p<0.001
Design-Build	p<0.001

Table 1g shows the mean differences in weighted performance evaluation scores between A + B contracting and lane rental, design-build, and traditional were all significant at the 0.001 level or lower.

For *unbonded concrete overlay projects,* the weighted performance preference rankings are as follows:

1st preference..... ...A + B Contracting
2nd preference.... ...Lane Rental
3rd preference..... ...Traditional
4th preference........Design-Build

TABLE 1h	A + B
Lane Rental	p<0.001
Traditional	p<0.001
Design-Build	p<0.001

Table 1h shows the mean differences in weighted performance evaluation scores between A + B contracting and lane rental, design-build, and traditional were all significant at the 0.0005 level or lower.

For *preservation projects with culvert replacement during 2-lane highway resurfacing*, the weighted performance preference rankings are as follows:

1st preference..... ....A + B Contracting
2nd preference.... ....Traditional
3rd preference..... ....Lane Rental
4th preference.........Design-Build

TABLE 1i	A + B
Traditional	$p<0.025$
Lane Rental	$p<0.025$
Design-Build	$p<0.025$

Table 1i shows the mean differences in weighted performance evaluation scores between A + B contracting and lane rental, design-build, and traditional were all significant at the 0.025 level or lower.

Open-Ended Questions. The state DOT construction engineers were also asked open-ended questions about the effective use of A + B contracting. The summary of issues discussed in these responses is as follows:[4]

Appropriate for projects with high Road User Costs	5 responses
Appropriate for projects with high Work Zone Congestion	5 responses
Appropriate for major projects	4 responses
Appropriate for urban projects	3 responses
Appropriate for projects with high ADT volume	3 responses
Appropriate for projects with high public visibility	2 responses
Appropriate for controlled access/interstate projects	1 response
Appropriate for bridge replacement	1 response
Appropriate for time constrained projects	1 response
Appropriate for multi-phased projects	1 response
Appropriate where contractor schedule input needed	1 response
Appropriate for projects with long detours	1 response
Inappropriate for projects with utility, ROW, soils issues	3 responses
Increases the ability to reward efficient contractors	3 responses
Suitable for all projects	1 response

[4] Not all respondents provided detailed answers to the open-ended questions. The responses listed represent a summary of all responses received from the open-ended questions.

In addition, most respondents indicated that they favored the inclusion of I/D clauses and several indicated that they set a maximum (B) time value that is intended to make the I/D clause meaningful and enforceable.

Case Study Research Methodology

Because of the stated preference for A + B contracting on each of the nine project types, phone interviews were conducted with three project managers from the Mn/DOT with experience in A + B projects. In addition, internal Mn/DOT documentation was reviewed to increase our understanding of some of the issues surrounding the use of A + B contracting. Information gained from the interview process was mostly qualitative in nature, while reviews of internal Mn/DOT documents yielded quantitative data.

The questions asked in the interviews were geared to address the rationale behind the weighting of each performance criteria. Specifically, the research team was interested in learning how the use of A + B contracting may have impacted the overall performance of the project. In addition to the performance criteria questions, two general questions were asked during the interviews to address perceptions about the strengths and weaknesses of A + B contracting. Interviewees were not necessarily expected to be able to comment on all areas of performance but those who were selected for interviews were able to speak to most of them.

Each interviewee was chosen because he or she had first hand experience in administration of an A + B contract and knowledge of the specific projects forming the basis of the case study. Interviewees for this case study were chosen after receiving input from Mn/DOT team leaders. Three interviews were conducted during March and April of 2005.

The internal document review examined four A + B projects performed by the Mn/DOT. Specifically, their projects were analyzed to determine project performance for comparison to a project of similar scope using design-bid-build contracting. The cost comparison examined first cost, final cost, bid durations, final durations, and approximate internal administrative costs as a function of total project cost. There was a "control project" of similar scope that utilized the traditional design-bid-build delivery. Each of the projects (three A + B and the control project) involved overlay projects on Interstate or trunk highways. Project data are noted in the following Table 2:

Table 2. A + B Comparison Project Data

	Bid cost/mile	Final cost/mile (without incentives)	Bid duration/mile	Final duration/mile	Percent internal cost
A + B Projects					
Project 96	$926,396/mile	$1,023,860/mile	9.35 days/mile	12.10 days/mile	9.9% of contract
Project 81	$926,009/mile	$949,344/mile	7.3 days/mile	7.62 days/mile	5.9% of contract
Project 41	$778,831/mile	$799,940/mile	8.13 days/mile	7.25 days/mile	6.8% of contract
Average of A + B	$877,078/mile	$924,381/mile	8.26 days/mile	8.96 days/mile	7.5% of contract
Design-Bid-Build comparison project	$815,442/mile	$923,032/mile	10.36 days/mile	10.14 days/mile	10.6% of contract

As can be seen in the table, the average first cost (bid amount) was slightly higher for A + B projects than for the traditionally procured project, but the final cost is approximately the same when incentive payments are eliminated. First costs were approximately 7.5 percent higher and final costs were less than 0.2 percent higher. Cost increases were 5.4 percent on average for A + B projects compared to 13.2 percent for the traditionally procured job. The similar final costs are the result of fewer cost additions during construction on the A + B projects. This reduction in cost growth may be caused by the need to have many of the potential conflict issues resolved prior to advertising for bids and/or starting construction.

The A + B projects were executed in a much more timely fashion, with 8.96 days per mile required for A + B projects compared to 10.36 days per mile for the traditionally procured project. A + B contracting appears to lead to approximate time savings of over 13 percent. However, the schedule extensions were more dramatic in A + B projects, averaging over 8.5 % while the comparison traditional project was completed in less time than allowed in the contract. Some of the Mn/DOT personnel commented that the presence of incentives for early completion makes contractors act more aggressively in pursuing delay claims, which may explain the greater schedule growth for A + B when compared to the traditional project.

Internal DOT administrative costs averaged 7.5% of contract amount for A + B contracts compared to 10.6% of contract amount for the traditionally procured project. This finding contradicts many of the suggestions in the literature that accelerated scheduling resulting from innovative contracting will create higher cost structures for state DOT's. We found no evidence to support that claim, and in fact found that internal agency costs are lower for A + B contracts. This is probably caused by the savings from a shorter overall duration offsetting the increases in overtime and shift premiums. Another contributing factor could be the additional preconstruction review, planning and coordination that occurs on A + B projects more than offsets the costs of redesign and claim review for traditional projects.

Interviews with the Mn/DOT project managers for the A + B projects revealed some differences of opinion regarding the use of A + B contracts. One project manager stated that the initial dollar amount of the bid (initial construction costs) would be slightly lower for A + B contracts than for traditional design-bid-build projects, one stated that first cost would be slightly higher, and one stated that they would be the same as for traditional projects[5]. All three project managers held the opinion that the final costs of construction, including change orders, would be similar for traditional delivery and for A + B. One project manager did say that contractors are very time sensitive when using A + B when there are I/D clauses, so it is important to have all environmental abatement, ROW acquisition, and utility relocation issues resolved prior to the start of construction. The contractors may be more aggressive in pursuing delay claims because a time extension may keep the incentives in play.

All three project managers indicated that use of A + B contracts added some complexity to the project management process. In particular, the scheduling of night work by the contractor places demands on department resources needed for inspections, field reviews, and quality control. Utility locations must be verified, conflicts resolved, and plans communicated to the contractor prior to the start of construction. Also, the need to negotiate for time in addition to money resulting from changes during construction adds another responsibility to project management. Lastly, one project manager interviewed stated an opinion that many bidders may not understand the award process since they had submitted the maximum "B" amount established by the state as part of their bid package.

Two project managers held the opinion that internal administrative costs would be higher for A + B projects because of the need to staff night-work and use overtime for department field personnel performing field reviews, inspections, and material testing. One project manager estimated that internal costs would be similar to other non A + B projects over the long-term, but could be higher during intense periods of work. This project manager stated that the range of internal administrative costs across all project types and contract methods for Minnesota DOT projects is 6%-8%, which compares favorably to the average for the A + B projects examined in the case study. The primary departmental operations affected by the use of A + B contracting are contract administration, because of the need for fast turnarounds on decisions and timely approvals, and field inspections, because of the need for some overtime and long shifts necessitated by night work and weekend shifts by contractor crews.

All three project managers agreed that Road User Costs are reduced by using A + B contracts, although one project manager stated that controlled access interstates have limited flexibility in reducing Road User Costs on a daily basis, but the overall duration of the impact is reduced. One project manager stated that

[5] One of the problems with comparing effectiveness of different innovative contracting methods arises from their use as both procurement AND delivery. Lane Rental and A + B contracting are innovative forms of procurement that use traditional delivery of independent contractor's and designers. Design-Build is an alternative delivery system with a single contract for both design and construction that can be procured in a variety of ways (low-cost, best-value, qualification-based, etc.) A complete discussion of the differences between procurement method and delivery method is beyond the scope of this paper.

impacts on third parties along the route (e.g. local businesses and civic entities) were reduced because of shorter project durations, but the other two stated no difference in third party impacts because access in maintained regardless of contracting method.

When asked if internal agency issues needed to be managed differently for A + B contracts, all three project managers indicated that some different approaches are required. Specifically, there needs to be a greater sense of urgency to resolve issues and more thorough plan review and coordination needs to take place prior to the start of construction when using A + B contracts. Related issues such as demolition, environmental remediation, and work sequencing need to be resolved much faster than on typical projects. In short, the department cannot afford to go in with as many "unknowns" as they might on a traditional project, and they must get supplemental agreements resolved quickly once construction begins.

The project managers indicated that A + B contracts are best suited for projects with high traffic volumes, projects with limited design complexity and those with few outside factors such as environmental issues, utility conflicts, etc. Unbounded overlays, bituminous overlays, grading projects, and mill and overlays under traffic are all projects that fit these descriptions. One project manager stated that he would suggest A + B contracting on any type of project, while another project manager stated that he would not recommend its use in rural areas where detours are available.

One project manager does not favor the use of incentive clauses because he believes contractors intentionally offer a "B" component on their bids that is longer than what they actually need, knowing they will be able to claim the bonus. Another anecdote regarding the use of I/D clauses involved district-wide policies. It was felt that if some projects had large I/D clauses, other projects without I/D in the district would suffer because contractors would commit resources to the projects where they could earn an incentive or avoid a disincentive. This could be problematic in rural, sparsely populated districts where few contractors are available for competitive bid work.

Conclusions

The findings of the research project confirmed some prior findings in the literature, but expanded or contradicted other prior conclusions. Prior research suggesting that the use of A + B contracts leads to shorter project durations was confirmed. The recommendation to minimize sources of outside conflict such as ROW acquisition, environmental issues, and utility conflicts was also confirmed in our study.

However, our findings appear to contradict some prior findings in the literature. Of particular interest is the result of a national survey of state DOT construction engineers which indicates that A + B contracting is the most effective means of contracting for each of the nine project types considered in this study. A + B contracting received the highest performance effectiveness score for each of the project types, and the differences in mean effectiveness scores were statistically significant in all comparisons of A + B to other contract types except for major corridor realignment/ expansion, where design-build mean effectiveness score was not statistically significantly different from A + B. This suggests that for all project types considered in this study, A + B contracts will create the greatest value when all relevant performance factors are considered. This appears to contradict the prevailing wisdom that the use of A + B contracting should be restricted to projects with high Road User Costs, traffic restrictions, or critical completion dates. State procurement protocols that do not allow for multi-attribute value consideration (e.g. those with statutory low-bid awards) will not capture the optimum effectiveness that innovative contract methods have to offer. From these results, it appears that A + B contracts should be considered for all projects, regardless of project type or critical performance factors.

One of the concerns in using A + B contracts is the belief that they will result in higher internal costs to the department. Our case study comparison did not find evidence of higher internal costs. In fact, the internal costs to the agency were actually lower on average for the three A + B projects than for the traditionally procured control project. Readers should note that the use of a single control project brings with it the bias of small sample size, but the average internal costs of 7.5% for the A + B projects investigated in this research fell within Mn/DOT target ranges (6-8%) for all projects state-wide. Lower than predicted internal costs may be witnessed because even though the intensity of work is high, creating a need for overtime and long shifts, the duration is shorter. The reduction in schedule may offset the increase in intensity. Another possible explanation for lower than anticipated internal costs is the extra time spent up-front to resolve potential conflicts and perform more thorough design and procurement reviews.

To summarize, it appears from the research that state DOT's should use A + B contracting on a wider variety of project types, and that internal cost premiums may be lower than indicated by prior research findings. Future research on internal cost comparisons for A + B contracting and other contracting methods using larger sample sizes will be required to confirm the findings of this preliminary study. However, results are promising for continued growth and expansion of the use of A + B contracting methods by state DOT's.

Acknowledgements

The authors are grateful to the Minnesota DOT for their financial support of this research project, as well as the cooperation of the Minnesota DOT personnel who assisted with data collection and interviews for this project.

References

AASHTO, (2001). *Primer on Contracting for the 21st Century: A Report of the Contract Administration Task Force of the AASHTO Subcommittee on Construction.* 4th Edition

Abi-Habib, N, Safir, A., Triplett, T. (2002). *NSAF Survey Methods and Data Reliability*, National Survey of American Families, Report No. 1.

Bolling, D.Y., and Holland, J. (2003). *Factors to Consider in Innovative Contracting*, Utah Technology Transfer Center.

Carpenter, B., Fekpe, E., and Gopalakrishna, D. (2003). *Performance-Based Contracting for the Highway Construction Industry*_Battelle: Columbus, Ohio

Federal Register, (2002). Vol. 67, No. 237; Tuesday, December 10, / Rules and Regulations.

Federal Highway Administration, (2002). *FHWA Briefing Special Experimental Projects No. 14*_ Innovative Contracting; February.

Florida DOT, (2000). "Office of Program Policy Analysis and Government Accountability." *Highway Construction and Engineering and Transportation System Maintenance Programs*, Chapter 4, Report 99-29, January.

Minnesota Department of Transportation, (2003). "Minnesota Statewide Transportation Plan" August.

National Cooperative Highway Research Program. (2001). "Guidelines for Warranty, Multi-Parameter, and Best Value Contracting", Report 451, Washington, DC

Ohio DOT, (2003). *Innovative Contracting Manual,* September 1.

Shr, J-F, Thompson, B.P., Russell, J.S., Ran, B., and Tsemg, H.P. (2000). "Determining minimum contract time for highway projects" *Transportation Research Record 1712*, Transportation Research Board, National Academies, pp. 1965-201.

Shr, J-F, Ran, B., and Sung, C.W. (2004). "Method to determine minimum contract bid for A + B + I/D highway projects" *Journal of Construction Engineering and Management*, ASCE, Vol. 130, (4), pp. 509-516.

South Dakota DOT, (1996). *Criteria and Guidelines for Innovative Contracting,* Office of Research Report SD95-07-X.

Guidelines for Quality-Based Contractor Qualification

R. Edward Minchin Jr.[1], M. ASCE and Gary R. Smith[2], M. ASCE

Abstract

The purpose of this paper is to propose a framework to qualify contractors for transportation construction projects using quality as one fundamental measure of the qualification process. Current qualification practices are briefly reviewed and examined for associated advantages and disadvantages. Quality performance measurement methods are then examined to determine related advantages and disadvantages with respect to inclusion in a quality-based qualification method. A revised framework of the traditional contracting process for departments of transportation is presented from examination of the literature and existing qualification systems. The application of the framework is examined with preliminary formulation of the resultant contractor qualification scores. Though the framework has many potential uses for a construction owner, the use of the scores is hypothetically examined for contractor capacity factors and as a pure qualification acceptance measure. The benefits of the proposed framework are that it is similar enough to existing methods to be familiar to users and uses existing data to lend a totally objective aspect to the model.

Key Words: Construction, bias, contractor, qualification, performance, bidding, implementation, quality-based, rating

Introduction

Quality-based qualification is a step soon to be taken in the evolution of public construction quality assurance. Shifting to quality-based qualification procedures could be a challenging move for agencies whose qualification practices follow traditional forms of prequalification and performance bonds. However, many departments of transportation (DOTs) have two driving forces for including quality performance as a key criterion in qualifying contractors. The first driving force is the continued erosion in the ability to staff projects. Reduced staffing includes those personnel traditionally assigned to quality control procedures. The second driving force comes from specifications designed for performance-based and performance-related contracts. With more contractors providing the project-level quality control function, the DOT's role has shifted to predominately a quality assurance role. Therefore, integrating quality performance measures into traditional qualification processes would be an obvious extension to current qualification practices.

Hybert (1996) identified five primary quality-based problems in major telecommunications contracts as follows:

---

[1] Assistant Professor of Civil Engineering, University of Florida, P.O. Box 116580 Gainesville, FL 32611-6580, minch@ufl.edu

[2] Dean, College of Engineering and Architecture, North Dakota State University, Engineering Building- Room 203, Fargo, ND 58105, Gary.Smith@ndsu.edu

1. Poor up-front definition of customer needs.
2. Incomplete evaluation criteria for awarding the contract (overemphasis on price).
3. Poor planning of project objectives.
4. Poor assimilation of necessary midstream project changes (driven by problems or improvements discovered during the project).
5. Measures and rewards driving the wrong performance.

Many of these same problems apply, in some fashion, to transportation construction contract administration. The key problem addressed in the remainder of this paper is the quality problem identified by Hybert as incomplete evaluation criteria for the award of the contract. Most transportation projects are awarded on the basis of price with award criteria generally being pre- or post-qualification, along with performance and payment bonds.

Thus, the position is taken that quality- and performance- related elements are missing in almost all current qualification processes. Also that these need to be addressed to ensure selection of a contractor who can deliver a contract at a cost and quality that is satisfactory to the owner. This paper will examine the status of qualification practices for public transportation construction, identify quality performance measurement methods, and propose a framework model combining prequalification and quality performance.

Qualification Processes

Thomas and Smith (1994) identified four strategies that DOTs use to qualify contractors:

- Pre-qualification
- Post-qualification
- Performance Bonds
- Contractor Licensing

Although the term "quality" is mentioned frequently as a potential attribute for contractor qualification systems, the literature search revealed no system currently being used that clearly linked quality to contractor qualification.

Pre-qualification, in very simple terms, is a review of information submitted by the contractor, allowing the qualifying agency to determine the contractor's eligibility to bid certain types of work. Contractor pre-qualification has been heavily documented in the literature. Post-qualification procedures have not been discussed as extensively in the literature, but information recovered suggests that most factors considered in post-qualification procedures are similar to those considered in pre-qualification. (Smith and Minchin 2001)

In addition to pre-qualification and post-qualification, Thomas and Smith (1994) reported that all DOTs require performance bonds. Even those who reported that they did not use some form of formal qualification, relied on the contractor's ability to provide a performance bond. Surety evaluations consider many of the same factors as DOT qualification systems. (Caby 2004)

Finally, because contractor licensing requirements and regulations vary widely, these are not considered to be an effective qualifying tool. Many licensing requirements are knowledge-based, but some are strictly a business tax revenue device and have little bearing on a contractor's ability to deliver a completed project.

Qualification Practices

The following list includes the predominant factors considered by DOTs (Minchin and Smith 2001) in their qualification procedures:

- Financial statement of the contractor including equipment inventory
- Years in business and years experience as a general contractor
- Summary level details on major projects completed (most require 3-5 years of data)
- Summary level details on projects in progress
- Experience of personnel relevant to the qualification request
- Company failure to complete projects
- Employee failure to complete projects
- Past disqualification or failure to be prequalified
- Work quality, based on post-project reports (usually qualitative, but limited use)

While each of these factors may be related to the contractor's ability to perform the work they are bidding on, only the last factor bears on the quality of work in place.

Pre- or post-qualification is not purported to be a cure-all. The system has advantages and disadvantages that have been identified in previous work as follows (Lower, 1982; Hauf, 1976; Nettleton, 1948, Russell, 1990, 1996):

Advantages

- Levels the playing field (on larger projects since large contractors may bid small projects but smaller contractors cannot usually bid large projects)
- Ensures competent, successful parties
- Eliminates unqualified parties, even though they can be bonded
- Significantly hastens bid award process and reduces the cost of bid solicitation
- Provides structure and discipline to the process
- Protects contractors from being awarded work they are incapable of doing
- Facilitates bidding by quality contractors who might have been inhibited from submitting a bid because of competition from unqualified bidders
- Allows more time for investigation of the contractors
- Decreases low-cost bias
- Reveals a contractor who may be unable to perform due to work backlog.

Disadvantages

- Additional screening creates workload for the DOT

- Qualified contractors may choose not to participate if the process is too extensive
- Developing, implementing, and evaluating objective criteria are costly
- Difficult to formalize process without introducing subjective judgment
- Potential for biased or erroneous denial of admission into bidding process exists
- Limits contractor's ability to expand into new areas in which contractor does not have prior experience.

The fact that contract and contractor failures continue to occur highlight the importance of considering other factors in the qualification process. Many DOTs collect subjective information by short questionnaires to evaluate performance as part of the contract close-out process. The information is most often not included in administrative procedures for limiting or disciplining contractors on the basis of their performance. If negative ratings affect the contractor at all, it is usually in the form of lower bid limits or higher retainage rates. (Minchin and Smith 2001)

Performance Bonds

Russell (1990) and Nettleton (1948) both suggest that using performance bonds alone does not guarantee that a contractor can perform the work. Bonds merely protect against financial losses but cannot predict delays, public inconvenience, or the level of cooperation between the contractor and the owner. However, there is no research that shows how pre-qualification predicts delays, eliminates public inconvenience, or guarantees that contractors can perform work. Surety companies can do a great deal to evaluate performance, but they lack the intimate knowledge of contractor site operations affecting quality in performance and project delivery.

Existing Or Proposed Quality Performance Measurements

Quality Performance Measurement System

The Quality Performance Measurement System (QPMS) evolved from early efforts at quantifying the cost of quality via rework. The customer focus section of the system contained a number of 'report card' feedback tools and assessment tools similar to those used by current DOT post-project evaluation questionnaires. The content outline from one such quality performance report card (CII 1989) is as follows:

- Understands and Translates Customer Needs into Requirements
- Manages Schedule
- Controls Cost
- Delivers Technical Quality
- Safety/Health/Environmental Issues
- Communications
- Work Process
- Administration

- Interpersonal Relations
- Leadership
- Subcontracts/Major Suppliers

Note that quality performance in this context is much broader in coverage. Technical quality is only a limited portion of the overall evaluation. (CII, 1989)

Certifications And External Ratings

Another method to ensure that only recognized 'quality contractors' work on DOT projects is to specify an external quality management system certification standard similar to certifications required for material testing laboratories. One such external or third party certification for quality management systems is ISO 9000. Another third-party approach is an audit process used internationally, called Construction Quality Assessment System or CONQUAS.

ISO 9000 One international quality standard has been developed and implemented through the International Standards Organization (ISO). The contractor, by checking quality and following formal inspection procedures, conducts the first level of quality assessment. The second level of assessment conducted is generally the assurance or inspections conducted by external parties (consultant inspections or possibly DOT inspections) who perform detailed examinations of the product or service provided. The ISO requirements mandate the third level of quality assessment be performed by an external examiner on the contractor's organization for quality. ISO certification is an expensive process for a contractor. (Corbett, et al 2002)

Construction Quality Assessment System (CONQUAS). The Construction Industry Development Board (CIDB) of Singapore developed this system of contractor assessment. It was developed as an objective quality measurement system for building construction that provides an incentive for encouraging contractors to improve the quality of their construction. The incentive process rewards a contractor by allowing a contractor with a high CONQUAS rating to bid higher than a non-rated or lower-rated contractor and still be awarded the contract.

Rather than performing a complete building inspection, the CONQUAS system is based on obtaining a representative sampling from each of the major work item classifications of building construction. The system first classifies the building by type, which changes the overall weights assigned to the work classifications. The three basic component areas are structural, architectural, and mechanical / electrical (M&E) work but each is weighted differently as shown in Table 1. The sample size is determined by the physical size of the structure. The sum of the three component scores is then used as the final building score.

An independent third party conducts the evaluation process (Prasertsintanah 1996). Recently, CONQUAS was introduced into the building industry of the United Kingdom. More details on CONQUAS are available in the literature. (CONQUAS 2003)

COMPONENT (1)	CATEGORY A (2)	CATEGORY B (3)	CATEGORY C (4)	CATEGORY D (5)
Building Types	Offices, shopping malls, industrial buildings, airports, hospitals, etc.	Condominiums, institutional buildings (schools, etc.)	Public Housing	Landed properties
Structural Works	30%	35%	45%	40%
Architectural Works	50%	55%	50%	55%
M&E Works	20%	10%	5%	5%
Total	100%	100%	100%	100%

Table 1. CONQUAS Building Categories and Weight Scheme

Questionnaires. Another method to obtain data related to a contractor's performance is through the use of questionnaires. Many DOTs currently use a survey or questionnaire form at the end of the project to evaluate the contractor's project performance. There may be as few as five or six general questions, or it may contain 30 or more detailed questions distributed over the same types of general categories. Surveys typically represent a one-time evaluation of the contractor's performance on a variety of issues like cooperation, timeliness of submissions, and others that represent quality from the viewpoint of the owner agency.

Model Development

It is within the context of these current processes and practices that the research team set out to develop a model that would measure the quality of a contractor's work as objectively as possible and do it in a way that can be used for qualification and bidding purposes.

As part of this research, the investigators conducted focus groups in four states. These focus groups were composed of DOT personnel as well as contractors and officials from contractor-advocacy organizations. The outcome of the focus groups was a list of factors that most impact the quality of a highway construction project. To determine which of these factors are most important, all state DOTs were sent a survey requesting that they rank the factors in order of importance.

Only four of 37 responding DOTs indicated that they were using any form of past performance rating in their qualification system. Many conduct post-project reviews, but do not have a formal process to integrate the information into their qualification-rating scheme.

A review of the literature revealed many distinct approaches to contractor performance measurements that could be modified to DOT quality assurance and qualification processes. Since many DOTs already use a post-project evaluation or contractor evaluation process to gather data on contractor quality, including this type of measurement scheme would not generate much resistance among those currently using surveys.

Research Approach. The research was conducted using the following process:

1. Literature Search.
2. Focus Groups. Focus Groups comprised of contractor personnel, DOT personnel, and contractor advocacy organization personnel were convened in Pennsylvania, Florida, Utah, and Kentucky. Participants were asked to help generate forms to be used as questionnaires for interim and end-of-project use. The focus groups were particularly useful in helping to identify the factors that most impact project quality.
3. Survey. Every DOT and the state contractor advocacy groups in the four states were sent surveys asking them to rank the eight factors listed most by the focus groups and also to add any new factors of their own.
4. Using this input, along with the experience of the research team, the top eight factors of construction quality were chosen and relative weights were assigned to the factors.
5. A questionnaire was devised for use as an interim report, another for an end-of- project report, and another as a state or district office report. These questionnaires contained questions dealing with each of the eight factors.
6. Projects were chosen in Florida and Pennsylvania and project personnel were asked to fill out end-of-project questionnaires on the projects, plus the district construction office was asked to rate the quality of each project. The project personnel also made all testing records on the project available to the research team.

Details of the Focus Group output can be found in the literature. (Minchin and Smith 2001; Minchin and Smith 2005)

Observations. Considering input from industry (focus groups and questionnaires) and the alternative approaches available for augmenting current processes, the following observations are provided:

- An ISO system would likely be beneficial to the construction community and to construction quality. However, there is no direct project feedback from ISO, only the periodic checks on company processes and quality management systems. The expense of maintaining registration and/or achieving registration has proven to be an economic barrier for some smaller contractors and represent a limit on competition. (Corbett et al 2002) Therefore, perhaps a program that requires the applicable elements of ISO 9000 without requiring certification should be considered.
- The third party project quality audit approach, similar to CONQUAS, has appeal from the fact that it is purely based on the quality of output. Unfortunately, it fails to measure critical elements of management and there is resistance from both the contractor community and DOT personnel to the third-party concept. (Minchin and Smith 2001) A scaled back version might

provide the product benefit without requiring a third party to conduct the project audits.

- There was no set definition of project quality developed. However, a list of factors that affect quality was agreed upon.
- Also, it was agreed upon that customer satisfaction is very important and cannot be measured in a purely objective, quantitative way. It is noted that all systems now used by DOTs utilize subjective data in rating contractors – in most cases to the exclusion of any objective data. This is why the objective aspect of the proposed model is so important.

A system that combines a questionnaire with some objective element of product measurement would cover both qualitative elements of quality along with quantitative elements.

Conceptual Model Elements

The following three elements are proposed for a quality-based qualification process: 1) A specific project performance report is required that records quality aspects of individual projects; 2) A company performance report that represents an aggregate picture of the contractor over time across all projects completed; and 3) a system that can accommodate the use of an agency's rating of a company's performance information to determine qualification.

The owner's (DOT) project field supervisor currently is responsible for most quality assurance activities on the project, including completion of the project evaluation report. Periodic evaluations could be developed that look at specific quality management characteristics throughout the execution of the project. Periodic evaluations are desirable because they reduce the possibility of bias toward contractor personnel (regardless of product quality) appearing in the final evaluation. The final project report would present the cumulative findings on the project. The interim and final reports could be integrated into the final model with a weighting scheme.

Individual project performance reports need to be combined either on a continuous basis or an end-of-year basis to assign the overall contractor evaluation or score. This score would be the basis for determining qualification or disqualification. Project evaluations can be combined based on the raw scores or weighted. The weights would likely be contract size (cost)-valued. While it is not likely that a quality contractor would produce more or less quality depending on contract size, larger contracts can be more complex and some agencies may feel that they deserve recognition in the overall consideration of a weight scheme.

Qualification considerations must be given to the time period covered or the number of projects included in the evaluation. If all the contracts are included, a contractor that completes 10 contracts per year could be compared or contrasted with those completing 2 or fewer per year. Another viewpoint is that over a long time period a contractor could establish a baseline that would permit poor quality projects for a period of time before the overall measure would begin to show degradation. A procedure for admitting new contractors to the bidding process will also need to be established. The time period or number of projects in the data set and the process for new contractor entry into the evaluations are matters of model implementation and are presented for consideration during model development.

Model Development

A system not drastically different from what most DOTs now have can be modified to capture a combination of objective project performance data from the project test results and subjective responses to the post-project questionnaire evaluations. The in-progress reports are used to measure the project on a frequent basis to ensure that quality performance issues are identified as soon as possible. In-progress reports also mitigate the effects of end-of-project bias. The post-project reports completed at the district or central office level are used to measure the contractor on contract issues and elements of the contracting process that a project engineer might not have exposure to in the course of the project. Data collected from the tests and inspections are also introduced into the evaluation process. Project performance reports are generated and combined with other project performance reports to create an overall quality factor for the contractor. Further details on the progress evaluation and the quality data schemes are discussed below.

An important element to evaluate in the final modeling phase is how many of these variables are needed for the intended purpose of trying to evaluate contractor quality performance as well as the quality and quantity of data available.

Table 2. Potential Elements For Project Performance Data Factors

ELEMENT	QUALITY FACTORS
Rigid Pavement	W/C ratio, consolidation/density, joint construction, air content, thickness, finish, strength, curing, and smoothness
Bituminous Pavement	Compaction density, asphalt content, gradation, surface smoothness, thickness, asphalt properties, aggregate quality, stability, void ratio, and skid resistance.
Bridge Deck	Consolidation/density, rebar cover, w/c ratio, density, mix temperature, curing, air content, strength, vibration/placement, smoothness, and form work
Structural Concrete	Consolidation/density, w/c ratio, curing, strength, air content, rebar placement/cover, vibration/placement, form work, and temperature control
Base Course	Compaction/density, gradation, thickness, moisture content, aggregate quality, uniformity, drainage
Embankment	Compaction/density, moisture, material quality, uniformity, drainage

Project Performance Factor Using Information from Questionnaire (PPF_q). Financial considerations are not related to contractor performance on the project. It could be argued from an opposing view, but many qualification systems are heavily attached to simply determining a capacity factor that will limit the overall amount of work the contractor can have in progress at any particular time. Knowledge about performance is more than financial data. After examining a number of project questionnaires currently used by DOTs and working with focus groups of contractor and DOT personnel in four states, several general content areas for the model were evident:

- Project management and control elements
- Project personnel
- Contractor organization
- Project plant and equipment
- Schedule adherence

The five factors retained for use in the Project Performance Factor (PPFq) based on questionnaire responses yielded the following equation:

$$PPF_q = a\,(\textit{Project Personnel}) + b\,(\textit{Project Management/Control}) + c\,(\textit{Contractor Organization}) + d\,(\textit{Plant and Equipment}) + e\,(\textit{Schedule Adherence}) \quad (1)$$

From the investigators' own experience, recommendations of industry (through the survey) and information collected from the original focus group meetings, the weights of a, b, c, d, and e were determined. They are 0.30, 0.20, 0.20, 0.10, and 0.20 respectively. (Minchin and Smith 2001)

The three questionnaires discussed earlier (interim, end, and district/central office) are then combined to derive the PPF_w, called the PPF_{wq} because each of the questionnaire results is weighted. The (weighted) PPF_{wq} would be calculated from a weighted summation as shown in Equation (2):

$$PPF_{wq} = I(PPF_q\,(\textit{Interim})) + E(PPF_q\,(\textit{End})) + D(PPF_q\,(\textit{District})) \quad (2)$$

Where I = Weight of Interim Questionnaires (Reports) Note: Scores from all Interim Reports are averaged and that value is multiplied by I
E = Weight of End-of-Job Report
D = Weight of District/Central Office Report

From the investigators' own experience, recommendations of industry (from the survey) and information collected from the original focus group meetings, the weights of I, E, and D were determined. They are 0.50, 0.30, and 0.20 respectively. (Minchin and Smith 2001)

Project Performance Factor Using Data (PPF_d). The unbiased nature of test results make them an attractive element for the model. These tests, performed on materials and workmanship, measure quality and lend an aspect of objectivity to the model.

The project's test information would be collected into a single factor called the project performance factor, data (PPF_d). The measure could collect all test results or only those of key quality elements and be recorded as a percentage of tests passing on the first attempt. Some projects have problems with mix designs and plant set-up and materials often fail initial tests. This is also true for soil tests and similar field measurements. Re-tests assure that the component meets acceptance criteria, but the contractor performance would only be measured on the initial test results.

Recent development of construction project software by American Association of State Highway and Transportation Officials (AASHTO) and commercial software developers give DOTs the opportunity to utilize comprehensive project management systems. These systems make the gathering of the test information required for this model easier than ever before. Even most states that don't have such a comprehensive system have some sort of information-system-based data warehouse that makes the gathering of the data relatively simple.

Project Performance Factor (PPF). When combined with the project performance factor, questionnaire (PPF_q), the final model would include both the qualitative response data and quantitative response data in a compatible form as seen in Equation (3).

$$PPF = m\ \mathrm{PPF}_{wq} + n\ \mathrm{PPF}_d, \tag{3}$$

Where m and n are overall weights for the two factors.

The Contractor Performance Factor requires a separate determination based on the results of the contractor's performance on a series of projects. In determining the values of m and n in Equation (3), the investigators used their own experience, recommendations of industry, and information collected from the original focus group meetings. Details of the process deriving the weights can be seen in the literature. (Minchin and Smith 2001) The equation for this phase of the research project was developed from this approach as shown in Equation 4:

$$PPF = 0.80(PPF_q\ (End)) + 0.20(PPF_d). \tag{4}$$

Contractor Performance Rating

The most challenging aspect of the model is how it will be used to rank, evaluate, classify, or code a contractor's performance. Once the field performance data are collected, and the appropriate score has been assigned, the contractor's performance must be determined. Assuming the PPF will be as described above, a contractor factor (CF) must be calculated to combine all the contractor's work into a single factor. Equations 5 and 6 may be used to derive an unweighted or project-cost-weighted contractor factor respectively.

An unweighted CF:

$$CF = \frac{\sum PPF}{n}, \qquad (5)$$

where n is the number of projects, or Cost Weighted (WCF)

$$WCF = \frac{\sum (PPF \times PV)}{\sum PV}, \qquad (6)$$

where PV is the individual project value.

Application Of The Model

To show that this model is practical and useable, DOT personnel were asked to use the model for one evaluation cycle (month) on one of their projects and the overwhelming majority reported that the use of the model did not appreciably add to their workload. (Minchin and Smith 2001) There are two ideas for the application of the contractor performance factor in the predominate methods of qualifying contractors. The first is to use the factor as a measure of the contractor's ability to bid work with a DOT and the second would be to modify the contractor's bid capacity based on the factor. If used strictly as a prequalification element this would leave traditional capacity factor determination processes intact. A minimum cut-off in Figure 1 (to be determined by the agency using it) would exclude a contractor from bidding. In the below-average range the contractor is basically on probation and must improve their rating in the next qualifying period or they could lose their right to bid work for a fixed time period, say 3-6 months. If they fall into the unacceptable region, their ability to bid work could be revoked for a period of perhaps 1-3 years. The specifics of the penalties for poor performers would need to be determined by individual DOT qualification personnel. It is not suggested that a normal distribution would definitely be the form that the population would take. That determination is beyond the scope of the research.

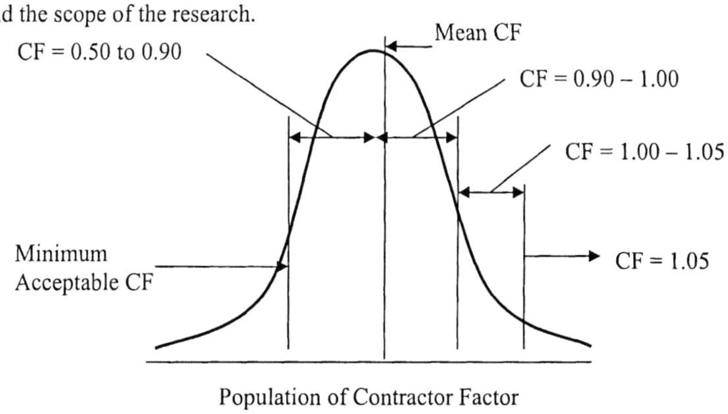

Figure 1. Use of Contractor Factor in Capacity Determination

Capacity Formula Application

Prequalification is the predominate system currently in use, but there are many variations for determining contractor work capacity. Where contractors are assigned a capacity factor, and their ability to bid work can be impaired or improved based on their work performance there is an opportunity to change the capacity formula by using the contractor rating. The contractor factor can be used to modify the contractor's work capacity rating. Work capacity ratings are often set on the basis of similar considerations. Use as a capacity factor multiplier could also be developed. Figure 1 illustrates the use of the total distribution of rating factors. The sample parameters and the assumed population parameters would both need rigorous statistical analysis before such a proposal would be viable. Note that this would enable a contractor to hypothetically increase its capacity rating if performance is consistently in the upper performance ranges.

The capacity adjustment would require contractors to strive for higher-than-average performance factors to prevent erosion of bidding capacity. This is not the same as a bidding preferential, but could have the same desired impact in the long term.

Other Model Applications

The two example applications, for qualification and bidding, are not the only ways one might use this quality-based factor. Other uses include:

- As a modifier in the bidding process to allow contractors with high aggregate performance factors to bid slightly higher (a preferential bidder bonus as found in CONQUAS)
- Quality Factor to be used to set penalty levels for substandard work
- Quality Factor to be used to set retainage rates. Higher quality would result in low or no retainage on DOT projects
- A Quality Index for determining when a contractor should be held in default or debarred in the middle of a qualification cycle
- As a method to determine special or emergency project 'short lists'
- An index to reduce or increase the size of bond required from the contractor

Summary

Fundamental to each potential use for a quality-based qualification system is the preliminary model of the factors to be used. The proposed project factor evaluation and contractor factor processes have been introduced. The major focus was on the evolution and considerations of various models and model components. The proposed integrated model represents implementation of concepts from pure questionnaire models, such as commonly used by DOT personnel, and product audit models into a single project rating. Data availability, sources, and 'workability' within most DOT qualification systems have been discussed. The data required for the proposed framework is available within existing data collection systems available to DOTs. The format and frequency of collection would need to be reconsidered. The proposed qualification process will require that contractors integrate quality processes into all phases of transportation construction to remain competitive in this highly competitive industry.

References

Caby, R., (2004) Surety Underwriter, Zurich, Personal Interview

Construction Industry Development Board (CIDB) (1998), *Construction Quality Assessment System (CONQUAS 21) Manual, 4th Edition*, Singapore

Construction Industry Institute (1989), The Construction Industry Institute Quality Management Task Force, "The Quality Performance Measurement System: A Blueprint for Implementation," *CII Publication 10-3*, The University of Austin Texas

CONQUAS (2003). http://collaborate.bre.co.uk/cqforum/public/conquas.htm

Corbett, C.J., Montes, M.J., Kirsch, D.A., Alvarez-Gil, M.J. (2002). "Does ISO 9000 Certification Pay?", *ISO Management Systems,* July-August, 2002, pp. 31-40

Federal Highway Administration (1979) Technical Advisory T5080

Hauf, H. D. (1976). *Building Contracts for Design and Construction*. John Wiley & Sons, New York, NY.

Hybert, P. (1996), "Five Ways to Improve the Contracting Process," *Quality Progress*, *V29 N2*, Feb 1996.

Lower, J.R. (1982). "Prequalifying Construction Contractors." *Management and Operations Journal AWWA*, May 1982.

Minchin, R.E. and Smith, G.R. (2001). "Quality-based Performance Rating of Contractors for Prequalification and Bidding Purposes," Final Report for NCHRP Project 10-54; Published as NCHRP Web Document 38, Washington, DC.

Minchin, R.E. and Smith, G.R ; (2005). "A Quality-based Contractor Rating Model for Qualification and Bidding Purposes," Journal of Management in Engineering, Volume 21, No. 1; American Society of Civil Engineers, Reston, VA, January, 2005, pp. 38-43

Nettleton, E.T. (1948). "Recent Experiences in Pre-Qualifying Bidders for Construction Projects." The Connecticut Society of Civil Engineers, Inc., 64th Annual Report. New Haven, Connecticut: Quinnipiack Press.

Prasertsintanah, P. (1996), "An Evaluation of a Premium Scheme in Tendering As Incentive Strategy, CIB W89 International Conference, Beijing, October 21-24, 1996.

Russell, J.S. (1990). "Journal of Professional Issues in Engineering Education and Practice." Vol. 116, no. 4, pp. 360-374. Illumina CSA

Russell, Jeffrey Scott (1996). *Constructor Prequalification, Choosing the Best Constructor and Avoiding Constructor Failure.* ASCE Press, New York, NY.

Schlickman, Jay, J. (1998), ISO 9000 Quality Management System Design, Optimal Design Rules for Documentation, Implementation and System Effectiveness, American Society for Quality, Quality Press

Thomas, H. R. and Smith, G. R. (1994). "Criteria for Qualifying Contractors for Bidding Purposes" Synthesis of Highway Practice 190, Transportation Research Board, National

Guidelines for Warranty Contracting for Highway Construction

Benjamin P. Thompson[1], MASCE, Stuart D. Anderson[2], MASCE,
Jeffrey S. Russell[3], MASCE, and Awad S. Hanna[4], MASCE

Abstract

This paper presents guidelines for implementing the construction warranty contracting method in the highway construction industry. A logical, step-by-step method for effectively applying the warranty contracting method for use in the highway construction industry is developed. Relevant issues that may present obstacles to the implementation of a warranty contracting program are addressed, while best practices compiled from an evaluation of current industry state-of-the-art that will assist an interested state highway agency in creating a warranty specification are also presented. In performing this research, the current use of warranty specifications among state highway agencies was examined, and an in-depth case study of the Wisconsin Department of Transportation's warranty program was conducted.

Introduction

State highway agencies (SHAs) are beginning to re-evaluate their goals and to shift more design and construction responsibilities to contractors, while focusing internal efforts more intensely on management and planning. This shift in responsibilities and, correspondingly, risks, may be facilitated by the use of innovative contracting methods, such as warranty contracting. SHAs are under increasing pressure to: (1) decrease costs and durations of highway projects; and (2) improve project quality and performance. These pressures are coupled with a trend towards organizational restructuring and reducing personnel. SHAs are also encountering evolving and expanding economies, changes in traveler demographics, population growth, increases in vehicle miles traveled, and changes in funding sources. Sources of SHA funding such as the gasoline tax and federal funding have become less certain, while at the same time competition for federal funding has risen, causing a greater uncertainty regarding budgets. In response, many SHAs are seeking more cost-effective, innovative methods for construction of transportation facilities and seeking to become performance-driven. Finally, SHAs are beginning to focus less on development and construction of new facilities and more on maintenance and rehabilitation of existing facilities (MDOT 1997). These changes may be dealt with, at least partially, by using methods that increase construction cost-

[1] Research Engineer, University of Wisconsin – 1415 Engineering Drive, Madison, WI 53706. phone: 608-263-9968, fax: 608-262-5199. bthompso@cae.wisc.edu

[2] Associate Professor, Texas A&M University – 3136 TAMUS, College Station, TX 77843. phone: 979-845-4414, fax: 979-845-6554. s-anderson5@tamu.edu

[3] Professor, University of Wisconsin - 1415 Engineering Drive, Madison, WI 53706. phone: 608-262-7244, fax: 608-265-9968. russell@engr.wisc.edu

[4] Associate Professor, University of Wisconsin – 1415 Engineering Drive, Madison, WI 53706. phone: 608-263-8903, fax: 608-265-9968. hanna@engr.wisc.edu

effectiveness by reducing life cycle costs of projects, reduce required agency resources, and encourage contractor innovation.

The contracting system currently used by SHAs, based on methods and materials specifications, involves specifying exactly what is to be built, how it is to be built, what materials are to be used, and how traffic is to be maintained during construction. This approach minimizes the performance risk to the private contractor building the project. SHAs are currently examining innovative contracting methods that shift more of the risk and responsibility for performance of the completed end product to the contractor. The contractor is seen as having the best control of the risks associated with quality and performance of the project. This paper is concerned with warranty contracting, defined as: "A type of performance–based specification that contains quality control requirements and guarantees the integrity of a product and the constructor's responsibility for the repair or replacement of deficiencies" (Wiseley 1998).

Warranty contracting has the potential to increase quality and improve performance, as well as reduce the amount of agency resources required to administer projects, while other methods may be useful for decreasing the delay to the traveling public caused by construction projects and for shortening project durations. This paper addresses the lack of guidelines for new users of the warranty contracting method.

Research Method

National Cooperative Highway Research Program (NCHRP) project 10-49 *Improved Contracting Methods for Highway Construction*, was conducted to develop guidelines for the application of innovative contracting methods in the highway construction industry. The research consisted of two distinct phases. Phase 1 focused on: (1) determining state-of-the-art of innovative contracting; (2) surveying highway construction industry for the use of warranties and multi-parameter bidding; (3) surveying other industry innovative contracting methods; (4) synthesizing literature and survey results; and (5) developing an interim report. The results and recommendations from Phase 1 embodied in the interim report were presented to a panel of industry professionals, and three innovative contracting methods were chosen for further development. Phase 2 of the project consisted of two steps: (1) developing selected contracting method(s); and (2) preparing the final guidelines. One of the methods selected for further development was warranty contracting.

At the beginning of Phase 2, a supplementary literature review was conducted to further understand the warranty contracting method. Additional data collection was accomplished through a survey of state DOTs currently implementing warranty contracting, as well as through an additional specification review. This information was used to create tables and a flowchart process model for the warranty method. The steps of the process model were developed and refined through an iterative process that resulted in the warranty contracting guidelines. In addition to the process model, a model warranty specification was developed for hot-mix asphalt cement (HMAC) pavement. This paper provides a general description of the process model developed for the warranty contracting method.

Model Description

Based on data collected from SHAs, as well as from studying individual specifications, programs, and projects, a process model was developed and refined for the guidelines presented in this paper. Explanations and descriptions for each step were gathered from SHAs that have implemented warranty contracting programs and were combined into a single guideline process model. The format selected to represent the guidelines was a flowchart. Figure 1 presents a generalized version of this flowchart.

Determine Motivation. The first step for the interested SHA is to determine the reasons they are interested in the warranty contracting method. Clearly articulating the objectives the agency hopes to accomplish through implementation of this method is crucial. Some objectives that may be considered include: improved performance of constructed product; augmentation of agency expertise; re-distribution of performance risk; and reduction of required agency design, testing, and inspection personnel. Even at this early stage, the agency must look at its available personnel and consider how the warranty contracting method may affect the personnel requirements for projects.

When considering the warranty contracting method, the SHA must take into consideration what end product(s) will be constructed using this method. The SHA should consider how best to achieve the objectives they have established. One factor to consider when choosing an end product is ease of implementation. The SHA must decide whether it has the necessary resources available to implement the warranty contracting method. Resources must be expended by the SHA at the beginning of the program to accomplish tasks such as drafting specifications; determining performance indicators and threshold values; educating contractors, sureties and other interested parties; and training their own personnel. A pilot process allows the SHA to gain valuable experience with implementing warranty contracting, and is thus recommended when implementing this or any other new, innovative contracting method.

Review and Understand "Best Practices". One of the most important steps in moving toward a warranty contracting program is to understand industry best practices for warranty contracting. A best practice is a method of accomplishing a function or process that is considered to be superior to all other known methods. An agency must gather information from other users of the warranty contracting method in order to understand the industry's best practices. In this step, the key is to understand and attempt to integrate best practices from other SHAs.

For more specific information, the agency should identify warranty contracting programs currently in use by SHAs, and contact them directly to obtain more information. Some of these states may have in-house reports evaluating their use of the warranty method, while others may be able to offer subjective or qualitative evaluation from personnel involved with projects using warranty contracting. SHAs most experienced with warranty programs for various project types are listed in Table 1.

Other agencies' successes are also an effective means to initiate change and build support for the warranty contracting method within an interested agency. To take advantage of these successes, an agency should collect "lessons learned" from SHAs who have implemented the warranty method. These lessons learned may be used to illustrate

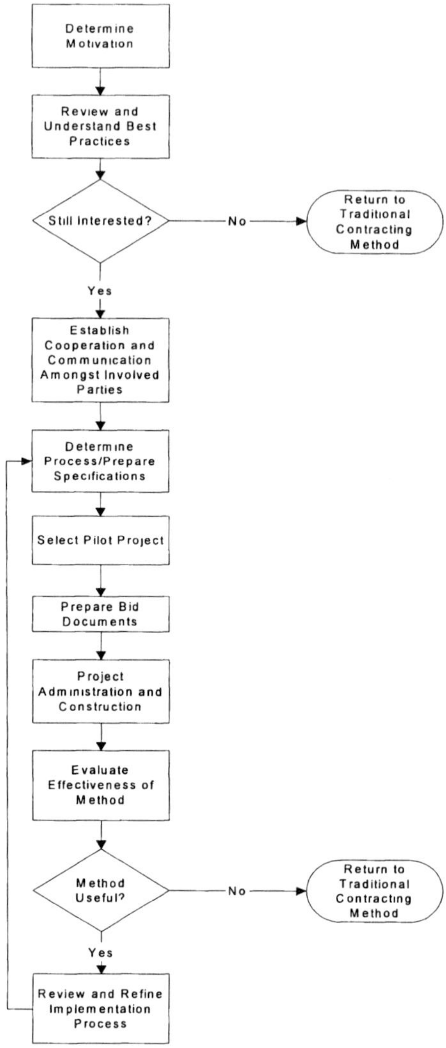

Figure 1. General Guidelines for Implementation of the Warranty Contracting Method (Thompson 1999)

possible improvements in existing systems, offer solutions to potential problems that may be encountered in the future implementation of the agency's own program, and indicate issues that remain unresolved and problems that require additional attention to solve.

Table 1. Use of Warranties (Thompson 1999)

State (1)	Significant Warranted End Product (2)
Indiana	Hot-Mix Asphalt Concrete (HMAC) Pavement
Maryland	Bridge Painting
Michigan	HMAC Pavement, Bridge Painting, Chip Sealing, Micro-Surfacing, Crack Treatment
North Carolina	Intelligent Transportation System Components
Pennsylvania	Pavement Marking
Wisconsin	HMAC Pavement, Portland Cement Concrete (PCC) Pavement

When attempting to understand industry best practices, a SHA should gather information relevant to evaluating project effectiveness based, in part, on the goals set earlier in the process. Some of the most common criteria used to evaluate effectiveness include cost savings, improved facility performance, and reductions in SHA personnel requirements.

Finally, when gathering information, the SHA must determine appropriate means of measuring each of these criteria.

Still Interested in Warranties? Early in the development of the warranty contracting program, an SHA must decide whether or not this method meets the needs and objectives set forth in the first step of this process. A comparison must be made between the information gathered by gaining an understanding of best practices and the objectives of the SHA. If the information gathered makes it evident that the objectives set are attainable using the warranty method, the SHA should continue to develop its program. If the objectives appear unattainable or could be better met by some other contracting method, the SHA should discontinue its investigation of the warranty contracting method and/or consider an alternative method. At this point, an important issue that must be considered is the feasibility of obtaining senior management support within the SHA central office and district offices for the use of the warranty contracting method. This is a critical factor in the eventual success of the program.

At this stage of the process, it would also be appropriate to identify the end product for which to implement warranties. In order to gain valuable experience implementing this contract method, the SHA should select a single end product to pilot the warranty process. Additional end products may be added subsequently as separate pilot programs. When selecting an end product, the SHA must consider how best to achieve the objectives established earlier in the process, as well as the ease of implementation, in terms of resources that must be expended to develop the process.

Establish Cooperation and Communication Between SHA, Contractors, Sureties, Material Suppliers, and Other Relevant Parties. It is important to bring all relevant parties into the warranty contracting method development process early to ensure "buy-in" of critical participants. Cooperation and communication between the SHA and the local contracting community are particularly essential. The first step in building cooperation and communication is to identify and contact all involved parties. Some of the major participants in the process include the local contracting community, surety community, and material suppliers.

Early contractor input is critical to developing the program. Contractor input is particularly essential when choosing criteria on which performance will be evaluated. The contracting community can offer valuable input as to which criteria they believe should be viewed as indicative of performance, as well as the threshold values they consider fair and reasonably achievable for any warranted product.

Surety companies are also very important to the process. Sureties must be willing to provide bonds for projects with warranties. If sureties have input in the decisions made regarding how warranty specifications are written, they will have a higher level of comfort with the entire process. Sureties will be more willing to provide the bonds if they have a thorough understanding of the bonding requirements and risks associated with the warranty program.

Material suppliers may also be interested in the development of a warranty contracting program. There may be changes in the types of materials or the types of quality control/quality assurance tests required by contractors to perform projects. Material producers may be able to offer valuable input regarding these issues.

Next, the SHA must institute a program to educate the involved parties, including their own personnel, regarding the changes required to implement the warranty program. Education of all involved parties is an important step in the development of a warranty program (Anderson and Russell 2001). SHA personnel may have a new role in the design and construction processes. For example, under a warranty contract, there is less responsibility during construction for SHA personnel and more for contractor personnel. On warranty projects construction inspection is not a primary function of SHA personnel. SHA personnel must allow the contractor latitude to perform work in a manner the contractor believes appropriate.

The contracting community also must be educated regarding their new roles, responsibilities, and risks. Contractors must be informed of the new requirements present under the warranty contracting method. The contractors may require the expertise to select appropriate materials and methods for construction, as these will not be specified by the SHA under the warranty contracting method. The contractor must understand how construction methods and techniques can affect the performance of the product. It is also important that contractors be informed of the historic ability of the end product to meet the performance threshold levels set. Contractors will be more likely to accept warranty contracts if the expected performance levels are readily obtainable. These historic performance levels can often be gathered from a SHA's pavement management database.

This historic ability of the product to meet the threshold levels is also important to the surety community. Surety companies need to be able to understand and quantify the risks involved with issuing a bond to a contractor for a project using the warranty contracting method. Documentation that the threshold levels are achievable, and that a contractor will be able to fulfill a warranty contract is important for sureties. Without such documentation, the bonds may be too expensive for contractors to afford. This is especially true over the extended period of a warranty contract, as compared to the length of a traditional contract (for example, five years versus one year).

Early discussions with contractors and sureties offering input into the process will help to facilitate communication and reassure both parties that the requirements of the program are fair, reasonable, and equitable. Early input will also enable any potential problems to be identified and addressed before the program is fully underway. A relationship among these parties must be established early to allow ongoing input from each involved party throughout the process.

Determine Process/Prepare Specifications. The first consideration when developing a warranty specification is compatibility with the state's standard specifications. Standard specifications are defined as a, "book of specifications approved for general application and repetitive use" (AASHTO 1998). Warranty specifications generally take two basic forms. The first is a set of supplemental specifications. These are defined as, "Approved additions and revisions to the Standard Specifications" (AASHTO 1998). Often, these supplemental specifications (or portions thereof) will become part of the standard specifications once the standard is updated and republished. Most states update their standard specifications periodically. The other possibility for a warranty specification is to take the form of a special provision, defined as, "Revisions to the Standard and Supplemental Specifications applicable to an individual project" (AASHTO 1998). When preparing warranty specifications it is important to keep in mind that they must be integrated into the state's existing standard specifications. When these documents do not fit together seamlessly, conflicts of interpretation may arise. Section 105.04 of the American Association of State Highway and Transportation Officials (AASHTO) Guide Specifications for Highway Construction (AASHTO 1998) gives an order of precedence for the standard specifications, supplemental specifications, and special provisions when these documents conflict. Table 2 lists and explains the elements to be included in a warranty specification (Wiseley 1998). This table may be used as a checklist for items that should be included in development of a warranty specification.

Draft specifications must undergo a rigorous, iterative revision process to resolve any inconsistencies or shortcomings that may present problems in the future. SHA personnel should thoroughly review the draft of the specifications, especially the personnel who will actually implement the warranty specification, both at the district and project site/field levels. Interested parties such as contractors and sureties should also review the specifications. Then, the draft specifications should be carefully revised and refined until all concerns have been addressed. Table 3 lists the major differences between traditional specifications and warranty specifications. This table may be used in conjunction with Table 2 to provide an outline of a warranty specification. The model specifications and guidelines developed through the research for this paper are available in NCHRP Report 451 (Anderson and Russell 2001).

Select Pilot Project(s). A SHA should not initiate implementation of warranties on all construction projects at once. Rather, the SHA should first target projects on which to pilot the warranty process to refine the program and gain experience through application of the specification. Some states use a single pilot project for an end product. Other states have selected as many as three pilot projects for a given product in a single construction season. Piloting may also be repeated in successive years for purposes of obtaining additional experience and performance data.

Table 2. Elements of a Warranty Specification (Wiseley 1998)

Item (1)	Explanation (2)
Description	• Describe what the specification covers and work required.
Warranty Length	• Establish length of the warranty. Can be fixed or varying using "A - B" system described in Step 11.
Bonding Requirements	• Establish the penal value of warranty bonds or retainage system. The penal value should be enough to cover the cost of remediating a worst case failure scenario. • Establish acceptable bond rating using A.M Best rating system. • Determine acceptable combinations of bonds (e.g., one full-length bond or a combination of bonds). • Determine steps to be taken if surety company falls below acceptable rating. • Determine penalties if contractor fails to renew warranty bond.
Maintenance	• Establish responsibility for maintenance activities and how they will be approved.
Conflict Resolution	• Determine composition of the Conflict Resolution Team (CRT). • Determine how and when CRT will be used. • Determine length of conflict resolution process.
Contractor Responsibilities	• Selection of materials and construction methods. • Securing of a warranty bond for the entire period of the warranty. • Guarantee of the end product for the entire period of the warranty. • Remedial action if any threshold levels are met or exceeded. • Establishment and submission of a Quality Control Plan and data. • Elective/preventative actions deemed necessary by the contractor. • Maintenance of third party liability insurance.
Department Responsibilities	• Approval of liability insurance and bond providers. • Determination of end product inspection method. • Annual inspection of end product. • Compilation and timely submission of an annual written report to contractor documenting performance of end product. • Notification of contractor regarding any required remedial actions. • Approval of materials and construction methods and techniques proposed to perform remedial actions. • Approval of any elective/preventative action performed by the contractor. • Specification of special requirements such as Quality Control Plans. • Definition of what constitutes an emergency condition. • Determination of responsibilities and acceptable response times in emergency conditions. • Determination of the existence of emergency conditions and remedy if necessary. • Definition and performance of Routine Maintenance (e.g., snow removal, sign maintenance, mowing grass) during the warranty period.
Performance Indicators	• Establishment of performance indicators and threshold levels. • Definition of conditions under which specified threshold values are not valid.
Requirements for Corrective Action	• Approval of remedial action and remedy period. • Establishment of activity types that void requirements for remedial action by the contractor (e.g., destructive testing procedures by the SHA or utility relocation).
Basis of Payment	• Establish measurement method for end product (i.e., per foot, meter, ton, square foot).
Insurance Coverage	• Examination of the general provisions contained in individual state statutes to determine coverage for all project participants. • SHA should have the contractor name the SHA as an additional insured party in their third-party general liability and automotive liability insurance policies.

Table 3. Differences Between Traditional and Warranty Specifications (Thompson 1999)

Attribute (1)	Traditional Specification (2)	Warranty Specification (3)
Award of Contract	Lowest responsible and responsive bidder.	Same as traditional specifications, unless adjusted by means of the "A-B" system.
Bonds	Payment and performance bonds at specified penal values must be executed and delivered to the owner prior to the start of construction.	An additional warranty bond at a specified value (usually less than the full contract amount) must be obtained from an approved surety company.
Maintenance	Contractor is responsible for and maintains the work during construction until final or partial acceptance of the work by the owner.	Contractor is responsible for maintaining the end product for the warranty period, including any necessary routine or emergency maintenance, as well as any required remedial actions.
Conflict Resolution Team	No Conflict Resolution Team.	Conflict Resolution Team (CRT) established for each project prior to the start of construction. CRT consists of 2 contractor representatives, 2 SHA representatives, and one mutually agreed upon third party representative, whose cost is shared equally by the SHA and the contractor.
Contractor Duties	• Maintain a quality control program. • Provide a copy of all quality control data to the SHA.	• Submit a quality control plan to the SHA. • Provide a copy of all quality control data to the SHA. • Perform any required remedial actions, preventive maintenance, and emergency maintenance necessary for the length of the warranty period.
Owner Duties	• Quality assurance testing. • Observing and sampling contractor-performed tests. • Monitoring control charts. • Directing contractor to take additional samples when necessary.	• Survey end product annually to measure performance throughout the warranty period. • Notify contractor of evaluation results and any remedial actions required within the prescribed time frame.
Measures of Construction Quality (measured at time of construction)	As stated in existing standard specifications. Quality indicators vary by end product and by state.	None.
Measures of Product Performance (measured at time increments during lifetime of facility)	None.	As listed in the warranty specification. Performance indicators will vary by end product, and by state, according to their own needs and experience.

Projects that have a high probability of performing well (i.e., low risk and low complexity) are the best candidates for a pilot project. For example, in the case of an HMAC pavement warranty, a pilot project should be selected that has an adequate subgrade. This approach should prevent factors not covered under the warranty from impacting the warranted performance of the pavement. By isolating the warranted performance from that of other, related factors, the SHA may also minimize the effort needed to determine responsibility for deficiencies that may occur in the warranted end product.

It is appropriate to have access to relevant traffic data such as the Average Annual Daily Traffic (AADT) so that loading conditions can be properly forecasted and performance can be measured against a standard. It may be inappropriate to measure a pavement with a high AADT against the same performance standards as a low AADT pavement. Thus, a project selected for piloting the warranty process should be one for which the SHA has access to appropriate data and information.

Prepare Bid Documents. In the case of a pilot project, some explanation of the warranty program should be included in the pre-bid process. While projects would generally be bid and awarded in the traditional manner, it is possible that they may be bid, awarded, or constructed in a manner differing from the traditional method. For example, the "A – B" system may be used to award the contract under a warranty specification. In this system, a contractor's bid is adjusted by some pre-determined and specified amount to reflect each additional year beyond a minimum number of years of warranty the contractor bids. For example, the minimum warranty period may be five years, and for every year beyond five the contractor receives a bid reduction of $20,000. If the contractor bids a seven-year warranty, the contractor's bid is reduced by $40,000 (i.e., (7-5) * $20,000) (Anderson and Russell 2001). Many of these differences would be specific to each project.

Project Administration and Construction. Most warranty projects will involve a change in responsibilities for the involved parties. Under a warranty contract, the contractor has more latitude to select materials, methods, and techniques than on a traditional project. The contractor is responsible for performing the quality control for the materials, as well as the quality assurance testing on the constructed product. The SHA may reserve the right to collect split samples of materials for informational purposes, but it does not need to execute a formal quality assurance program. The SHA, however, needs to document all costs associated with project management and construction for use in comparing the cost of warranty projects with traditional projects.

Under the traditional system of specifications the SHA would prescribe the mix design, construction methods and techniques, quality assurance testing, method of payment, and the incentives/disincentives to be used on a project. The SHA may also specify the quality control tests to be performed on the project by the contractor. Under a warranty specification, the contractor has the freedom to determine these items, although the SHA may specify minimum requirements for materials and quality control plans. The only criterion that determines compliance with the warranty specification is the performance of the product relative to the established performance criteria.

Evaluate Effectiveness of Method. Once the warranty period has been completed, the entire project must be evaluated. This is an important step in evaluating the effectiveness of warranties within a SHA. Some important items to evaluate in relation to the original objectives set by the SHA include: (1) performance of the final constructed project; (2) level of design and testing/inspection personnel required for the project; (3) use of SHA and outside expertise; (4) risk distribution; (5) amount of claims and litigation, if any; and (6) total cost (construction plus SHA management).

Costs must be compiled in order to evaluate the life cycle cost effectiveness of the warranty project. In order to make a valid cost comparison with projects bid in the traditional manner, the SHA should keep track of the unit costs bid by the selected contractor, as well as the unsuccessful contractors on warranty projects. These costs can then be compared to costs of the traditional contracting method to evaluate the effectiveness of the warranty project. Table 4 offers a list of some of the items that should be included, both for standard contracts and warranty contracts, when compiling costs to evaluate the life cycle cost effectiveness of a warranty project.

Table 4. Costs for Comparison Between Warranty and Traditional Contracts (Shober et al. 1998)

Costs Included in Standard Contracts (1)	Costs Included in Warranty Contracts (2)
Mixture bid price	Asphalt Pavement, Warranted bid price
Asphalt bid price	Costs of training and use of conflict resolution team
Tack coat bid price	State delivery costs (reduced from standard)
Quality management bid price	Extra distress surveys and reports
State delivery costs	Extra tests for disputes, traffic counts, etc.
State maintenance costs for length of warranty	
Conflict resolution (found to be negligible)	

Method Useful? At this point in the process the SHA must determine whether the pilot project (or projects) has met the objectives set for it. To do this, the results of the pilot project must be compared with the objectives established by the SHA. Using the information gathered in the previous step, the SHA must determine whether the warranty pilot project was cost effective. This may not be an easy task at this point in the process. However, if implementation of warranty contracting is to be continued, an attempt at cost effectiveness analysis must be made. The Wisconsin DOT's (WisDOT's) "Three-Year Progress Report" on asphaltic pavement warranties is an excellent example of the procedure that should be employed to determine cost effectiveness (Shober et al. 1998). However, WisDOT's analysis is preliminary, and must be continued at least until the warranties expire. At this point, feedback on the program should be solicited from contractors, sureties, and other interested parties, as well as from district personnel. Lessons learned should be captured and shared with all involved parties.

Review and Refine Implementation Process. To continue implementation of the warranty program, the process must be evaluated and, where appropriate, modified to include additional best practices and lessons learned as they become available. Feedback is essential to improve the process. Any problems that were encountered on the pilot project(s) must be identified, and appropriate corrective measures taken. These recommended changes should be incorporated into subsequent warranty specifications. As the SHA gains more experience with warranty contracting, the specifications can be refined to more effectively accomplish the SHA's program goals. Specifically, the threshold values established for each performance indicator must be carefully evaluated. It is important that the threshold values be sufficiently stringent to keep performance indicators at the desired level, but at the same time reasonable and achievable for the contractor.

It is important to establish a formal feedback procedure to improve the efficiency and effectiveness of the warranty method. The recommended changes must be incorporated into the warranty program through revision of the special specifications or provisions. It is desirable for the SHA to continue this iterative process of improvement throughout the life of the warranty program, though these iterations are especially important for the pilot phase. As the program is refined, special attention should be paid to determining the types of projects on which it is appropriate to use warranties.

Once the warranty program has been modified and the specifications have been revised, the SHA should consider implementing warranties on additional projects. The warranty program should be implemented again using the revised supplemental specifications or special provisions on a second project. Warranties may be applied to more projects as the SHA becomes more comfortable with the method. The warranty method may also be expanded to warrant the performance of additional end products. The pilot development and implementation process should be repeated for each newly selected end product.

The SHA must continue to gather cost and performance data on warranty projects and on the process in general. Information on life cycle costs should be collected and updated constantly to improve the effectiveness of the warranty program. Warranties are not the appropriate contract administration method for every project. However, the warranty method can be effective if applied correctly under the appropriate conditions in a structured process as described in this paper and represented in Figure 1.

Critical Success Factors

All of the steps listed and described above are considered important for implementation of a warranty contracting program. However, some of the most critical factors affecting the degree of success achieved through the implementation of warranties will be described more fully below.

Training. All parties involved in the contracting process must understand their roles and responsibilities under any new contracting system. SHA responsibilities will be changed, as will contractor responsibilities. Most notably, with the warranty contracting method, the SHA no longer performs quality assurance inspection and testing, and no longer specifies methods and materials. In addition, the contractor must formulate, submit, and perform a quality control program. It is important that the SHA employ personnel with the appropriate expertise to implement a warranty program.

Appropriateness of Warranty Method for Projects. It is very important that an alternative contracting method be appropriate for the projects on which it is used. Both the SHA and the contractor must possess appropriate levels of resources to execute the roles defined in the training process above. Projects chosen for application of the warranty contracting method must be compatible with the objectives set by the SHA for implementing that method. Each SHA must define criteria to make clear which projects are candidates for a warranty contract. These criteria can only be defined based on experience with the method on multiple projects.

Communication. Communication among major parties involved in the warranty contracting program is crucial throughout program development, as well as throughout the entire duration of individual projects. It is important for all involved parties to understand and commit to the program early in the development process. SHA personnel, contractors, sureties, and other involved parties must bring their expectations into alignment early in the process, and must continue to clearly communicate their concerns and feedback to each other to improve the process. A working relationship among all involved parties must evolve and continue throughout the duration of each project. Communication between the involved parties must be used to provide a means for resolving conflicts quickly during a specific project. Partnering is a practice that may help achieve this relationship.

Integration of Design, Construction Methods and Techniques, and Sequences of Work. Warranty contracting must be chosen as the contracting method early in the life of the project. SHAs interviewed in the course of this research felt very strongly that, to be effective, a warranty must be designed into a project from the beginning. The scope of work, design of the project, and preparation of specifications and contract documents are all elements of the project that may be affected by the choice of contracting method. It is highly desirable for a warranty process to be designed into a project before these elements are prepared. Design considerations, selection of construction methods and techniques, and sequences of work are all interrelated and will be impacted by the choice of contracting method.

More Up-Front Investment by SHA. The interested SHA must be willing to initially invest resources in the development of the warranty contracting method. These resources will be invested in program planning, specification development, data collection for establishment of performance indicators and threshold values, training of SHA personnel to implement the warranty program, and education of involved parties. As the warranty process matures, these costs will be recouped as life cycle costs of individual projects are reduced.

Support of SHA Upper Management and Industry Buy-in. Without the approval and support of senior level management within the SHA and the buy-in of local industry personnel, a warranty contracting program cannot be implemented successfully. It is important to obtain support from these groups early by soliciting their input in the process development.

Project Selection Criteria. One of the most important concepts related to the development and implementation of the warranty contracting method is that of project selection criteria. No single contracting method is appropriate for all projects. Rather, a contracting method must be chosen that is appropriate for the specific conditions of each project. The chosen method must also meet the general objectives of the SHA. This paper does not advocate the replacement of the traditional low-bid system. However, when applied in a rational manner under the appropriate circumstances, the warranty contracting method may be useful and effective.

Case Study

The Wisconsin Department of Transportation (WisDOT) has been using warranty specifications for asphalt cement pavement since 1994. The WisDOT warranty program was studied in detail and examined for this research. WisDOT has traditionally been involved in highway construction from conceptual design through operation and maintenance. The WisDOT asphaltic concrete (AC) pavement specifications moved from a traditional method and materials specification to Quality Control/Quality Assurance (QC/QA) specifications in the late 1980's. This QC/QA program became very comprehensive by the mid-1990's, and WisDOT felt that the next logical step in specification evolution was a move to warranty specifications. The warranty specification development process began in 1994. Since its inception, the WisDOT warranty program has included nine asphalt paving projects, and has expanded into Portland cement concrete projects. In 1994, when the warranty program was initiated, WisDOT formed a joint team with the Wisconsin Asphalt Pavement Association and the Wisconsin Division Office of the Federal Highway Administration to draft an AC pavement warranty specification.

The project selection criteria used by WisDOT focused on the experimental nature of warranty contracting. Factors considered in an attempt to minimize complexity and eliminate as many variables from the performance evaluation as possible included location, consistent foundations, and relatively constant levels of existing distress over the length of the project. One of WisDOT's objectives in implementing warranty contracting was to reduce its inspection and quality assurance personnel demands. This decrease was understood to be coupled with an increase in personnel requirements for annual pavement evaluations. WisDOT found that contractors on the warranty pilot projects tended to test the same or less than on a traditional project. On one project, State Highway 11, WisDOT found that the contractor's frequency of quality control (QC) tests was about 30% less than what WisDOT would have required. This illustrates one of the potential benefits of warranties. Not only is the state's duplicative testing eliminated, but the amount of testing performed by the contractor may also be decreased (Shober et al. 1998).

WisDOT views the warranty specification as a means of paying the contractor to take a specified, reasonable risk. The warranty specification allows the contractor to select materials, mix designs, techniques, and a quality control program, while holding them responsible for pavement performance for five years. It was important to WisDOT that the risks assumed by contractors were reasonable. The warranty contracting method has the advantage that it fits within the traditional "low-bid" system, although the number of bidders on warranty projects may be decreased compared to traditional methods-based specification projects.

Under the WisDOT warranty specification, the warranty bond for a project must be in effect for the entire five-year period of the warranty to insure that any required remedial actions be completed. The warranty bond covers only the AC pavement in the amount of the estimated cost of a thin overlay. WisDOT attempted to keep the bond's penal value reasonable, thereby enabling the contractors to obtain the bonds from their sureties, and to prevent contractors from exceeding their bonding capacity.

The quality of the WisDOT warranty projects has not been diminished from the quality of a traditional project. None of the distress thresholds have been exceeded on any of the warranted pavements. Two values, the International Roughness Index (IRI) and the Pavement Distress Index (PDI) give a composite view of the quality performance of a pavement. Average IRI and PDI values for both warranted and non-warranted pavements are listed in Table 5. All the warranted pavements have lower IRI (International Roughness Index) and PDI (Pavement Distress Index) values than a typical, non-warranted pavement of the same age(Shober et. al. 1998).

Table 5. AC Pavement Performance: Warranted vs. Non-Warranted (Shober et al. 1998)

Performance Indicator	Pavement Age			
	New	1 Year	2 Years	3 Years
(1)	(2)	(3)	(4)	(5)
State Average IRI – Non-Warranted	1.00	1.12	1.29	1.36
Average IRI – Warranted	0.80	0.83	0.79	0.80
State Average PDI – Non-Warranted	0	4	11	18
Average PDI - Warranted	0	2	5	8

Summary and Recommendations

In an attempt to reduce the amount of SHA resources required on highway projects, reallocate performance risk, increase contractor innovation, increase the quality of constructed products, and ultimately to reduce life cycle costs of highway projects, warranty contracting is becoming more common. Generalized guidelines to assist an inexperienced user in implementing a warranty contracting program have been described in this paper. While these guidelines were designed for SHAs having little or no warranty contracting experience, they may also be of value to SHAs with more developed programs as a means of formalizing a program, or as an additional source of information for determining current industry best practices regarding warranty contracting.

Several important issues to include in an implementation plan for the warranty contracting method have been identified. It is important to examine each of the critical success factors summarized above when implementing a warranty contracting program. Training of involved parties regarding changes in their roles and responsibilities under the new contracting method is essential. The appropriateness of the method must be examined on an individual, project-by-project basis in terms of resource availability as well as method applicability and potential of the project for achieving agency objectives. Communication throughout the development of the program and the duration of each individual project is a major factor in the success or failure of a warranty contracting program. A working relationship among all involved parties must evolve early in the program and continue throughout the duration of each project. Finally, it is important to choose the warranty contracting method early in the development of a project, and to cultivate support and commitment within the senior-level management of the SHA and the local contracting community.

Acknowledgments

The authors express our appreciation to the National Cooperative Highway Research Program, Transportation Research Board, under the National Academy of Sciences for sponsoring this research. Publication of this article does not necessarily indicate acceptance by the Academy of its contents, either inferred or specially expressed herein. The authors would also like to thank the NCHRP Project 10-49 Panel for their comments and direction during the research process. Finally, the authors would like to thank Byron Blaschke at the Texas Transportation Institute, and Robert Smith, attorney at Wickwire Gavin in Madison, WI for their insight, input and hard work on this project.

References

American Association of State Highway and Transportation Officials (AASHTO). (1998). *Guide Specifications for Highway Construction.* Washington, D.C.

Anderson, S.D. and Russell, J.S. (1999). "Guidelines for Warranty, Multi-Parameter, and Best Value Contracting." National Cooperative Highway Research Program (NCHRP) Report 451. National Research Council, Washington, D.C.

Michigan Department of Transportation. (1997). "1997 Business Plan." http://www.mdot.state.mi.us/acrobatfiles/businessplan.pdf. Lansing, MI.

Shober, Stephen F., Mark Woltmann, Bill Duckert, Scott Schwandt, John Volker, Tom Fudaly, and Gerry Waelti. (1998). "Asphaltic Pavement Warranties Wisconsin Department of Transportation Three-Year Progress Report." Wisconsin Department of Transportation, Federal Highway Administration, and Wisconsin Asphalt Paving Association, Madison, WI.

Thompson, Benjamin P. (1999). "Guidelines for the Implementation of Warranty and Best Value Contracting in the Highway Construction Industry." Master's Thesis. University of Wisconsin, Madison, WI.

Wiseley, Patrick W. (1998). "Current Use of Warranties in Highway Construction." Master's Thesis. University of Wisconsin, Madison, WI.

Incentive/Disincentive Contracting Practices for Transportation Projects

David N. Sillars, P.E., M.ASCE[1] and Jean Pol Armijos Leray[2]

Abstract

Timely highway construction has an increasingly significant importance. As the cost of highway improvements increases and the ability to secure new rights-of-way decreases, the nation's highways are becoming increasingly congested. Highway construction now must be accomplished within existing rights-of-way and invariably creates inconvenience and cost to the traveling public. Reducing the duration of this inconvenience and thereby reducing public cost is a key target for many state highway agencies (SHAs).

Several alternative delivery methods have been used for timely delivery of projects. One of these methods is Incentive/ Disincentive (I/D) Contracting. State Highway Agencies need an effective process to implement I/D contracting in their projects; the purpose of this research is to develop a simplified process for implementing I/D contracting in State Highway Agencies, using current, general guidelines.

Incentive/ Disincentive (I/D) contracting experience exists among various state departments of transportation. Reportedly, this alternative contracting process often occurs in reaction to the needs of individual projects; guidelines used for the creation of I/D contract parameters are under development, not fully established, and are often instituted ad-hoc rather than through a generally-applied set of principles. This research reviews and consolidates many sources of published guidance regarding the use of I/Ds in highway contracting and presents a holistic set of broad principles of project-specific I/D contracting; the purpose herein is to provide a necessary overview toward understanding how each step in the process fits within the whole. This work may be used by various departments of transportation in the U.S. or other agencies that may wish to utilize I/D contracting and generally is compatible with guidance suggested by the Federal Highway Administration (FHWA). This work is intended to complement existing, more detailed discussions presented in the references.

Background

The purpose of this work is to consolidate many sources of published guidance and recent research regarding the use of incentive/ disincentive contracting in the United States. This work is intended to be a useful synthesis of such practices, targeted as its audience practitioners among the various state departments of transportation and

[1] Associate Professor, Director R.C. Wilson Graduate Program, Department of Civil, Construction and Environmental Engineering, 202 Apperson Hall Oregon State University, Corvallis, Oregon, 97330.

[2] Project Engineer, PCL Construction Services, Inc., Central Link Light Rail C755, 15426 35th Avenue South, Tukwila, WA 98188

other transportation agencies. Researchers may also find this work useful in understanding gaps that remain in the development of this style of contract procurement and management.

Introduction and Definition

SHAs have historically used the traditional Design-Bid-Build contracting method to deliver highway projects. In this well-known method, the contract is awarded to the lowest responsible bidder, which results in a focus on lower costs for the public sector but does not emphasize quality or timely delivery in the project. There are alternative methods of contracting available to State Highway Agencies that take into consideration the increasing demand for quality and timely delivery; Incentive/ Disincentive (I/D) contracting is one of these methods.

From 1927 until 1984 the Federal Highway Administration adhered to a policy that prohibited the use of bonus payments for early completion. The policy was based on the belief that "the FHWA should not have to pay 'extra' just to have a project completed early."

In the 1980s, increased highway demands and varying on-time construction performance created interest in innovative means to better encourage on-time or accelerated performance. In the early 1980s, the National Experimental and Evaluation Program (NEEP) Project #24 was conducted; this program showed that I/D provisions were a "valuable, cost-effective construction tool". In 1989, the FHWA issued a technical advisory, "Incentive/ Disincentive (I/D) for early completion - T5080.10" (FHWA 1989), the purpose of which was to "provide guidance for the development and administration of incentive/ disincentive provisions for early completion on highway construction projects or designated phase(s)". An incentive/ disincentive provision was defined in the Advisory as "a contract provision which compensates the contractor a certain amount of money for each day identified critical work is completed ahead of schedule and assesses a deduction for each day the contractor overruns the I/D time."

According to the 1989 FHWA technical advisory, the approval of I/D provisions for early completion was considered appropriate only for critical projects or phases of projects where traffic inconveniences and delays must be minimized. It was expected that this would result in significant savings or benefits for the users of highway construction projects. The responsibility of selection of these projects was left to each state.

European Study Tours

The Office of International Programs of the FHWA has regularly organized teams whose mission is to visit international transportation departments and discover practices that may be beneficially adopted in the U.S. The teams include representatives from the FHWA, the American Association of State Highway Transportation Officials (AASHTO), the Transportation Research Board (TRB) and other organizations. Of the many such studies conducted, at least three study tours have discussed innovative construction procurement and contracting practices in selected European countries and suggest delivery methods applicable to the U.S. highway agencies.

The European Asphalt Study Tour (1991) found that lane rental contracting methods including incentive/ disincentive clauses were frequently used in the United Kingdom on critical projects that must be opened at the earliest possible time. The lane rental with incentive/ disincentive concept was only seen as practical if a large degree of certainty existed on the scope of work. If several structures were involved or if completion depended on completion of an adjoining section by another contractor, the concept was not used. Officials stated that a change order on a lane rental project is essentially a blank check (FHWA 1991).

The Contract Administration Techniques for Quality Enhancement Study Tour (CATQEST) (FHWA 1994) found that incentive and disincentive clauses were seldom used for exceeding the specified minimum quality level, but some contracts included bonuses for early completion. Incentives (and disincentives) were considered unnecessary since most contracts in the European Union were procured under a best-quality system. The best-quality procurement system relies on selection of contractors based on prior performance metrics gained through historical experience in performing projects with the agency; cost is not an overriding consideration in such a system. In this instance, it was expected that the successful contractor provides its best effort.

In 2004, a team reviewed construction management techniques in Canada and various western European countries. Where incentive/ disincentive techniques were used, they were employed primarily to relieve traffic congestion; however, the use of incentives and disincentives were infrequent and quite varied. Incentives for timely completion included making lump-sum payments only at completion of specified segments of work, congestion lane pricing, and somewhat infrequent use of liquidated damages (FHWA 2005).

TRB Task Force A2T51

In 1988, the Transportation Research Board (TRB) established the Task Force on Innovative Contracting Practices (A2T51). This task force was created to identify and suggest methods to improve promising innovative contracting practices (Van Ness 1990).

One of the areas of study addressed by the task force was the evaluation of the effects incentives/ disincentives have on bidding procedures for transportation projects. It recommended that specifications should include incentive and disincentive provisions to encourage better quality and early completion of critical projects. Another recommendation was that the A+B bidding concept should be considered for wider implementation. The task force's findings were documented in Transportation Research Circular Number 386, entitled, "Innovative Contracting Practices" (Transportation Research Board 1991).

Special Experimental Project No. 14 (SEP 14)

The TRB Task Force A2T51 requested that the FHWA establish a project to evaluate some of the task force's specific recommendations. The FHWA responded with creation of the Special Experimental Project No. 14 (SEP 14). Since SEP 14 began in 1990, several innovative contracting practices have been evaluated. SEP 14 provided the State DOTs with a vehicle for assessing various types of non-traditional contracting methods on Federal-aid highway projects.

Four innovative contracting techniques were originally proposed, used and evaluated by a number of SHAs. These four techniques were among those showing promise in studies such as those conducted by the TRB Task Force (as documented in Transportation Research Circular No. 386, "Innovative Contracting Practices") and by the 1991 European Asphalt Study Tour report (Weseman 1995). The techniques were:

- "Cost-plus-time (A+B) bidding,
- Lane rental,
- Design/build contracting, and
- Warranty clauses."

By 1995, after a five-year evaluation period, three of the four experimental techniques originally identified were declared operational (A+B bidding, lane rental, and warranty clauses) and were no longer considered to be experimental in nature. Based on this designation, State Highway Agencies may use these techniques without FHWA Headquarters approval on Federal aid projects. In 2002, the FHWA published a Final Rule regarding design-build in the Federal Register (FHWA 2002). This rule, promulgated as a result of the legislation authorizing the Transportation Equity Act for the 21st Century (TEA-21), adds design-build as a contracting method for projects receiving Federal aid.

The bidding and contracting technique that is of interest for the purpose of this paper is the (A+B) method because it has been often accompanied by incentive/disincentive provisions, known as *A+B+I/D*. A+B bidding is discussed below in the section *on Selection of Contract Type.*

NCHRP Guidelines for Multi-Parameter Contracting

NCHRP Project 10-49, Improved Contracting Methods for Highway Construction Projects, was initiated in January 1997 and documented in 2001 in the NCHRP Report 451—Guidelines for Warranties, Multi-parameter, and Best Value Contracting (Anderson and Russell 2001). The scope of this NCHRP report for multi-parameter bidding was to provide comprehensive guidelines and develop a general approach or methodology to implement a multi-parameter contract method called *A+B+I/D+Q*. The use of incentive and disincentive provisions in A+B contracts was studied in Chapter 3 of Report 451 of the NCHRP program under the name of "Multi-Parameter Bidding and Contracting". The entire report expands the *A+B+I/D* method through the addition of a quality parameter *Q*. The *Q* parameter was described as a measurable product-related (not management-related) parameter. Criteria for selecting a Q parameter was provided in the report.

I/D Contracting Implementation process

A summary process for implementing I/D construction contracting is proposed in this research in Figure 1 and discussed further in this paper. It outlines the steps that a State Highway Agency (SHA) or others may take to implement I/D contracting and is similar in format to (and borrows from) a process model proposed in the NCHRP Report 451 (Anderson and Russell 2001). The different phases of the project life cycle are covered in the context of this flowchart. The general implementation of I/D contracting is described by the steps and decisions found in the flowchart. It is important that the current level of experience of the SHA with I/D contracting is assessed; according to this assessment the users may decide whether to follow all the steps described in the flowchart, or to tailor the process to better meet the capabilities and policies of the agency.

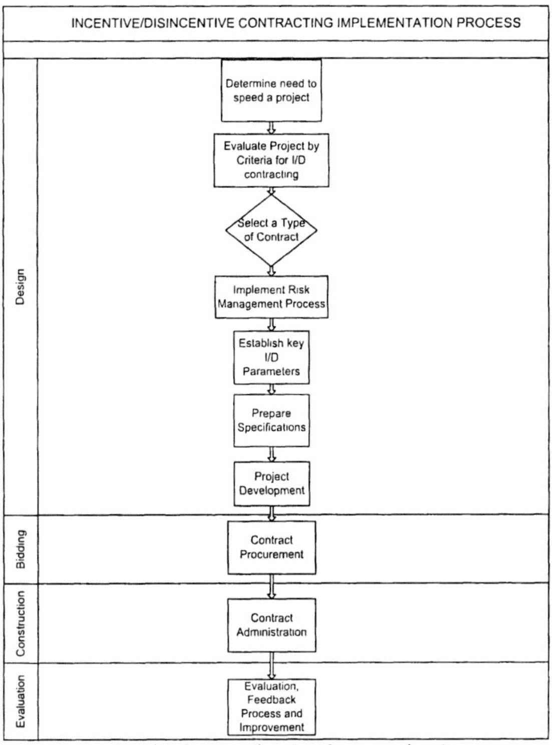

Figure 1: I/D Contracting Implementation Process

There are two general approaches for implementing I/D contracting in State Highway Agencies. The first defines, from a portfolio or list of potential projects, which projects are advisable to use for I/D contracting. This method sets strategic direction and planning for the state highway agency in a more centralized environment. Portfolio effects to consider include the effect of causing a contractor to favor an incentivized project over a non-incentivized project, and the potential for multiple incentives to unexpectedly aggregate in a given period, causing short-term financial burden to agencies. These items are touched on briefly in this paper, but in general unexplored in the literature. For further explanation about portfolio management, the reader is encouraged to review the NCHRP Guidelines for Multi-parameter contracting referenced at the end of this paper (Anderson and Russell 2001). The second approach, presented in this paper, defines the implementation of I/D contracting on a decentralized project by project basis. This method recognizes the uniqueness of projects and evaluates, for a particular project, its suitability for I/D contracting and development of key contractual parameters.

For an agency that lacks experience in I/D contracting, an individual pilot project may serve as the starting point for evaluation of an I/D implementation process. As I/D contracting is used on more and more projects, the feedback obtained from evaluations would improve the implementation process of this alternative method of delivery.

The implementation process includes (see Figure 1):

1. The identification of goals and needs for accelerating a project schedule;
2. Evaluation of the likelihood of success in applying I/D methods to the project;
3. Selection of the type of contract;
4. Implementation of a risk management process;
5. Determination of key parameters;
6. Preparation of specifications;
7. Project development;
8. Contractor procurement;
9. Contract administration; and
10. Evaluation process.

For simplicity, the term "I/D" is used in this paper to encompass two major types of incentive/ disincentive contracts. The two types of I/D contracting are termed herein as "I/D only" and "A+B+ I/D", and are distinguished where appropriate. The distinction between these two methods is discussed below in the *Selection of Contract Type* section.

Need for application of time-critical methods

The first step of the I/D contracting process is to identify the external need and motivation for acceleration or speeding up the delivery process of a certain project.

The determination of utilizing I/D for a particular project should occur during the early phases of project development. In this manner, the adequate allocation of resources to design and coordinate the project may be better planned.

Five potential objectives that SHAs may consider when deciding if a project needs to be accelerated are summarized below from the NCHRP (2001) Guidelines for Multi-parameter contracting:

1. Shorten Project Duration: The number one reason to implement I/D contracting is to reduce the time it takes to complete the construction of highway projects beyond what has historically been achieved with traditional highway contracting;
2. Reduce Inconvenience to the Traveling Public: Shortened project durations translate into reduced levels of inconvenience to road users, minimized congestion, and rerouting of traffic occurs for shorter periods of time;
3. Lessen Potential Impact on Local Businesses and Communities: Shortened project durations reduce the time that business access is potentially disturbed. On some I/D projects, a contract segment upon which I/D values are to be applied may consist of a roadway that restricts access to local businesses and neighborhoods;
4. Encourage Innovative Construction Processes: Innovative construction processes and methods are sometimes needed to accelerate the project. The contractor could also be innovative with the sequencing of construction in relation to traffic throughout the duration of the project;
5. Improved Perception of Construction on the Public: The use of I/D provisions will decrease the time required for construction. This means that the traveling public will observe that progress is made on a consistent and continuing basis. Agencies have been subjected to public criticism because of what appears to be long time periods where contractors are not seen working on the project consistently.

The majority of projects that use I/D provisions are completed on time or sooner, which support the use of I/D provisions (Arditi et al. 1997). However, savings in contract durations are often accompanied by increased project costs when compared to projects without I/D provisions. Clearly identifying the objectives the agency desires to achieve through the alternative contracting process achieves two goals: first, the use of the contracting process is more easily and readily defended under conditions of public criticism, and second, a baseline is developed for evaluation of the success of the project (and other projects) (Anderson and Russell 2001).

I/D Project Success Criteria

After the need for using the alternative contracting method of I/D for fast delivery of the project has been established, the next step of the I/D contracting process is to evaluate if the project itself meets the criteria for successful implementation as an I/D project. One of the greatest challenges involved with the implementation of an I/D contracting process is determining which projects are most effectively administered with an I/D contract. The criteria below must also support any other agency-specific objectives that may apply, such as administrative regulations for alternative contracting (legal review, etc.) or budget constraints.

The 1989 Technical advisory (FHWA 1989) recommends that I/D provisions should not be used routinely, but only for completion of projects or portions of projects for which one or more pre-selected criteria exist. Several sources have developed criteria for evaluating whether a project is a candidate for the use of I/D contracting techniques (Anderson and Russell 2001; FHWA 1989; Utah Local Technical Assistance Program (LTAP) 2005). These lists contain many similarities, yet each provides unique perspective as well; a consolidation follows:

Public Inconvenience

1. Where high-traffic volumes exist, typically in urban areas;
2. Where traffic restrictions, lane closures, or detours result in high road user costs (RUC);
3. Major reconstruction or rehabilitation on an existing facility that will severely disrupt traffic;
4. Lengthy detours of high traffic volumes;

Constructability

5. The agency seeks contractors' expertise to facilitate an earlier completion;
6. The I/D phase(s) can be completed in one construction season or less;
7. Traffic control phasing can be structured to maximize a contractor's ability to reduce the duration of construction;
8. The project is relatively free of utility conflicts, design uncertainties, or right-of- way issues which may impact the bid letting date or the critical project schedule;
9. Availability of contractors with sufficient resources;

Public Safety

10. Safety concerns during construction, including impacts to public, pedestrian, and/or worker safety;
11. Disruption of emergency services;
12. Emergency response to an unexpected loss of highway facility;

Public Priority

13. It is in the public interest to complete the project as soon as possible, or by a specific completion date;
14. Adjacent neighborhoods or businesses would suffer a significant impact;
15. Project will complete a gap in the highway system;
16. Major bridges out of service;

17. Completion time constraint;
18. Interference with major public events; and
19. Highly sensitive project (businesses impacted, political issues).

The relative importance of the criteria needs to be determined in order to make a decision. The criteria could be ranked in order of importance. The criteria could be weighted to reflect level of importance, perhaps by survey or expert panel. Lacking a set of standardized guidelines, the relative importance of these and other criteria is determined individually by the SHAs, each of which may emphasize the criteria differently. In recognition of this situation, the determination of the relative importance of these and other criteria is anticipated to be project specific.

A project does not need to necessarily meet more than one of the aforementioned criteria to be considered for I/D contracting; however, many states require that the project meet or exceed a certain RUC value to use I/D contracting. The agency must determine how to use the criteria to make a decision as to whether the project is a good candidate for I/D contracting. The following are some possible decision rules described by the FHWA (1989) :

- "Require the project to meet all criteria;
- Require the project to meet the top three criteria;
- Require the project to meet a minimum RUC and three of the other criteria; and
- Require the project to meet a minimum RUC, have utilities and ROW cleared, and two of the other criteria."

Research through the Utah DOT (Utah Local Technical Assistance Program (LTAP) 2005) has reported that some agencies specify a minimum threshold RUC level such as $2,000 - $3,000 per day.

Not every project is right for the use of I/D contracting. The agency should select projects that will maximize the chances for successful completion using I/D. If the project meets or exceeds the criteria for applying I/D set by the agency, the contract type should be selected and later a risk management process should be implemented. If the project does not meet the criteria for using I/D as determined by the agency, another established contracting method should be considered.

Selection of Contract Type

Once the project has met the criteria for becoming an I/D project, the next step is to decide what type of I/D contract the project should use.

Incentive/ Disincentive provisions have been used in two different alternative methods of delivery: *I/D only* and *A+B+I/D*. Each of these methods is discussed below, and each has been found to be successful in reducing construction time to some degree (Herbsman and Glagola 1995). An understanding of both the advantages and disadvantages of each I/D method is important so that the agency can decide if the selected contracting method can be effectively used on their highway construction projects.

I/D only. The *I/D only* method of contracting has been under experimentation for a long time. It is the typical low bid method with the inclusion of I/D clauses. The contract time, defined by the SHA, is part of the bid documents. This is a contracting system in which a contractor is motivated to accelerate the construction after the bid has been awarded. For each day that the contractor finishes the construction ahead of schedule, an incentive fee will be paid, but a penalty will be charged for late finish. The contracting agency specifies the time required for critical work and uses this provision for those critical segments where traffic inconvenience and delays should be minimized. This method has been used in numerous occasions by State Highway Agencies; however, many practitioners are not very pleased with past results. These practitioners argue that since the contract time is determined by the SHA, the contract time requirements may not be very accurate and may be determined on the high side; therefore, contractors would easily be able to save some contract time and earn an incentive without making special effort. Research by Herbsman et al. (1995) showed that under this procurement method, most of the contractors received an incentive fee, and that it has been very rare for anyone to pay a disincentive fee.

A+B+(I/D). One of the techniques that is of more current interest is the cost-plus-time bidding (A+B) method accompanied by the application of incentive /disincentive provisions, indicated herein as *A+B+(I/D)*.

This method demands that bidders propose time of completion and a proposed cost; both values are used in the low bid determination, but only the cost value is used to determine the base contract amount. Under this method, each bid submitted consists of two parts:

- The cost or "*A*" component (essentially, the traditional bid for the contract items), which is the dollar amount for all work to be performed under the contract; and
- The time or "*B*" component, which is the total number of calendar days proposed by the bidder to complete the project (or perhaps a significant segment of the work) (Weseman 1995).

The successful proposer is determined by developing an "equivalent" bid value, based on a combination of the bid for the contract items and the associated cost of the time. The lowest and best bid may be determined by the SHA according to the formula shown in Equation 1, where *Cost/day* is commonly the calculated RUC or other agreed value. Where I/D provisions are included in the contract, the I/D value may sometimes be used as the *Cost/day*.

A+B Equivalence Formula: A + [(B) x Cost/day] (1)

The FHWA recommends that an I/D provision be used with an A+B contract, rather than with straight competitive bidding where contract time is defined by the

SHA. The I/D amount must be substantial enough to motivate the contractor to achieve an earlier completion date or discourage the contractor from finishing later than the contract completion date. Without the I/D provision, the contractor has little reason to finish any earlier than the time bid to secure the contract. The contractor's estimate for the completion of critical work becomes the contract time. For critical projects that have high road user delay impacts, the *A+B+I/D* bidding method can be an effective technique to significantly reduce these impacts.

Under this method, the contractor is motivated in two ways. First, through the inclusion of time in the competitive bidding, the contractor is incentivized to propose timely completion durations; these durations often are significantly less than the engineer's estimate. In addition, after the award, the contractor is motivated to reduce construction time to earn an additional incentive amount. The possibility that contractors may inflate their durations to earn substantial incentives when the project is awarded can be overcome by stipulating in a contract clause that if the contractor exceeds the engineer's time estimate the bid can be rejected. Of course, since time is used to determine award, excessive time bids would additionally put a bidder at a competitive disadvantage.

For *A+B+I/D* projects, prior to the bid opening at which calendar days will be received, the agency should determine the method needed to convert the time bid into a dollar value using concepts such as Road User Costs (RUCs). Table 1 provides a comparison among key parameters between these two major contract types. The agency should determine under which project conditions each type of I/ D contracting is most effectively applied.

Table 1. Comparison of I/D contract type parameters

	Incentive/ Disincentive Only (I/D Only)	Cost-plus-time-plus-I/D (A+B+I/D)
Contract Time	E	B
Contract Amount	A	A
Bid Comparison Amount	A	A + B*s
Final Contract Amount	A +ΔE*s	A + ΔB*s

Note—any adjustments to final amount due to changes are ignored to simplify comparisons.

Once the contract type has been determined, a comparison must be made between the information gathered on the project characteristics and the objectives established previously. If it is evident from the information gathered that the objectives established are attainable using I/D contracting, the SHA should continue pursuing the selected I/D method. If the objectives appear unrealistic, the SHA should stop the implementation of the selected I/D method for this project and choose another contracting method (Anderson and Russell 2001).

The remaining steps on the flowchart in Figure 1 may be applied to either type of I/D contracting. Special emphasis in this paper has been placed on the *A+B+I/D* method; however, the *I/D Only* method is a variation that requires little adjustment from the techniques discussed herein.

Lane Rental. Although usually discussed as a separate contracting technique, Lane Rental is a method that has significant similarities to I/D contracting. In the United Kingdom, the original form of I/D was the lane rental charge. In this method, the bidding contractor estimates the time needed for lane closure and includes this cost (based on lane rental fees set by the agency) into the bid. The lowest bidder is determined by the lowest bid for the cost of work items, including the cost of lane rental (Herbsman and Glagola 1995). If the contractor overruns the time estimate, an additional rental fee will be deducted, or charged, to the contractor and if ahead of schedule, an incentive of refunded rental charges is received for work completed early. The main advantage of this method is that it allows the contractor to choose the best work patterns (day, night, weekend, detour, etc.), since rental fees may be time-specific (often on an hourly basis). Lane Rental has evolved into a very similar method to *A+B+I/D*, in which case the incentive/disincentive daily rate is equal to the rental fee rate for lane closure (Herbsman and Glagola 1995).

Risk Management Process

It is important to evaluate agency resources and risks when using the I/D contracting method for a certain project. An efficient risk management process to identify, assess, allocate, respond to and control risks is needed. Allocation of risks should be done using the basic principle that the party best able to control or manage a risk should take responsibility for that risk. If contractors perceive high externally-caused risks that are not adequately addressed, they may increase their bid proposals to take into account potential, difficult-to-predict costs.

SHA contract administration and other risks are created under I/D contracting. Claims may result from poor or untimely information flow from the agency. Appropriate levels of resources, including personnel, equipment, and materials, must be available to both the SHA and the contractor to accomplish an I/D project. Increased frequency of inspection and testing are likely in I/D projects; poor management may compromise the quality of the delivered project (Anderson and Russell 2001).

The impact on agency personnel requirements should be assessed. On an I/D project, the total number of agency workhours (outside of that required to manage potential claims) is approximately the same as would be required on a traditionally contracted project. However, the agency personnel may be impacted substantially when the contractor works extended hours per day and/or per week. The number of agency workhours required per day and/or per week may increase substantially because of the extended work schedules that are required to meet a reduction in project duration. The increase in agency workhours per day translates into more agency personnel required at a given time for the project, even though the personnel are required for a shorter length of time due to the reduced project duration. As reported by the FHWA (Anderson and Russell 2001), "one agency estimated that approximately 2½ times the traditional number of agency personnel would be required at a particular point in the project if the contractor works 7 days per week, 24 hours per day."

Field personnel (inspectors, testing personnel, and construction engineers, etc.) must have their work schedules carefully pre-planned to avoid disruption. Although the agency may reduce extended workhours by limiting the number of hours per day and per week, the result may be less effectiveness of the I/D contract method (Anderson and Russell 2001).

The assignment of personnel who are familiar with or are trained in I/D methods will reduce the likelihood of failure due to poor agency response. These personnel include planners, designers, field engineers, inspectors, and supervisors who must become comfortable with a shift of decision-making responsibilities toward the field. The need for timely responses during construction will require a change in the process of decision-making. (Anderson and Russell 2001).

Incentive/Disincentive Parameters

Once the agency has determined the motivation for using I/D contracting or what they want to achieve through the use of I/D in the project, key contractual parameters must be determined. These include the determination of road user costs (RUCs), I/D amount, I/D caps, maximum time allowed and minimum time allowed for the incentivized segment. Due to the infrequent use of I/D contracting methods and a lack of formally defined methodologies, strong engineering judgment, coupled with well-kept historical data become important factors in developing effective I/D parameters.

Incentive/Disincentive Time. The incentive/ disincentive time was defined in the 1989 Technical Advisory document as "the time (calendar days or completion date) established for the contractor to complete critical work on identified roadway(s) and/or structure(s)." As previously explained in the section related to types of contract, the party defining the I/D time would vary depending on whether it is an *I/D only* or *A+B+I/D* contract. If it is an *I/D only* contract, the SHA defines the I/D time and if it is an *A+B+I/D* contract, it is the contractor who defines the time (FHWA 1989).

A thorough analysis of the construction schedule is required to determine I/D time. The key to successful implementation of I/D time is an accurate evaluation of the time required to complete the incentivized segment, based on regular work hours and normal efficiencies (Shr and Chen 2004). Underestimates of time may cause unreasonably high bid prices and unattainable acceleration. Overestimates of time set by the SHA may allow the contractor to earn the maximum amount without making an increased effort. This would also penalize the public since the I/D phase(s) may not be completed as soon as possible but will cost more due to the incentive payment (FHWA 2005).

Normal construction time is generally based on a competent contractor working five days a week, eight hours a day, while an accelerated time should be based on the performance of a good contractor working extended or extra shifts with additional workers for six or seven days a week. Continuous seven-day workweeks should be avoided since extended periods of work without days off can result in high turnover rates for contractor and inspection personnel.

In addition, maximum and minimum project durations may be defined. Typically, the maximum allowable duration is specified for the project. This requires the use of a scheduling tool; a critical path method (CPM) schedule, with production rates, should be used on complex projects. Historical production rates can be used to arrive at a traditional schedule by estimating an 8-hour/5-day workweek. Standard production rates should then be converted to reflect an accelerated schedule using engineering judgment and experience.

A few agencies also specify minimum project durations using CPM scheduling techniques. The agency can use the same scheduling tools to adjust for accelerated production rates to arrive at an estimated shortened schedule. By establishing a minimum project (or segment) duration, the agency may avoid the receipt of bids wherein the bidder strategically proposes a high construction cost, but also proposes an unrealistically short construction time in order to secure the bid. Contractors should be required to submit a CPM schedule with their bid. (Anderson and Russell 2001).

Since production rates are not normally tracked by SHAs, some agencies may find it difficult to establish reliable acceleration time-frames to put forward in the bidding documents. Systems used for establishing accelerated timeframes include Oregon DOT's use of time/motion studies to establish baseline production rates, which are then modified to evaluate the effect of increased shifts and added work-days, and the sophisticated software program, CA4PRS (Lee et al. 2004), used on several CALTRANS projects. Other methods for establishing baseline production rates include widely-available standard production rates provided through national sources such as R.S. Means and F.W. Dodge. Standardized procedures should be established for determining acceleration targets, and minimum and maximum schedules; such formalized procedures will provide a basis for addressing public comment as well as setting a standard from which future improvement may be made with project experience.

In an *A+B+I/D* contract , if the I/D time is considered for awarding the contract (the time required to complete the incentivized segment), the I/D time bid would be determined by the competitive environment and would reflect the most likely durations for the project as seen from the perspective of the contractors.

The use of calendar days or completion date (rather than working days) has been shown to be effective contract parameters. Working days are subject to interpretation and the incentive increases the pressure for non-working days to be declared by the project engineer. This leads to an increased adversarial relationship among the parties during the construction phase (FHWA 2005).

The FHWA recommends that the season of the year in which the project will be constructed should also be considered in determining the I/D time for calendar day projects. The project should be such that an I/D phase can be completed in one construction season. Weather days and legal holidays should be included for calendar day projects.

Total Project I/D Time. A total project incentive/disincentive contract assumes that the I/D amount is based on the schedule of the entire project. This incentive type should be used only when the entire project is causing impact to traffic during the full duration of the project; otherwise, the contractor may shorten a non-impact portion of the project and gain rewards without substantial benefit to the public.

Segmentally Applied I/D Time. Using segmental I/Ds, the I/D amount and I/D time are defined for only certain portions or phases of the project that are causing impact to traffic. Examples may include a bridge in a larger project, a roadway segment that is particularly congested, an interchange, etc. Segmentally applied I/D times are recommended.

Incentive/Disincentive Amount. It is very important to determine the appropriate dollar amount per day for I/D provisions for early completion of projects. To be effective and accomplish the objectives of I/D provisions, the FHWA Technical Advisory recommends that the dollar amount be of sufficient benefit to the contractor to encourage his/her interest, stimulate innovative ideas, and increase the profitability of meeting tight schedules. If the incentive payment is not sufficient to cover the contractor's cost for the extra work (additional crews, overtime, additional equipment, etc.), then there is little incentive to accelerate production, and the I/D provisions will not produce the intended results. The manner in which the I/D amount is determined and documented is very important to the contracting agency in case it becomes involved in a legal dispute with the contractor (Jaraiedi et al. 1995).

Following are some of the guidelines of the 1989 Technical Advisory for determining the I/D amount. According to the Advisory, a daily I/D amount should be calculated for a particular project based on:

1. "Established construction engineering inspection costs;
2. State related traffic control and maintenance costs;
3. Detour costs; and
4. Road user costs."

The value of the road user cost is predetermined by the contracting agency and, if it becomes the basis of the I/D or is used for determination of the *B* portion of the bid, it is specified in the proposal. Costs attributed to disruption of adjacent businesses should not be included in the daily I/D amount. Engineering judgment may be used to adjust the calculated daily amount downward (not upward) to a final daily I/D amount that:

1. "Provides a favorable benefit/cost ratio to the traveling public where the cost is the daily I/D amount and the benefit is the calculated daily savings in road user and State Highway Agency costs; and
2. Is large enough to motivate the contractor. If a favorable benefit/cost ratio cannot be realized and/or the resulting daily amount is not high enough to motivate a contractor, the project should not be further developed as an I/D project" (FHWA 1989).

Most agencies have a standard method for estimating RUC on traditional projects. If none exists, a method for determining RUC should be developed. The agency should decide how much of the estimated RUC will be included in the contract for the time-related costs. For incentives, agencies have used anywhere from 10 percent of the estimated RUC to the entire amount of RUC in previous $A+B+I/D$ contracts. Typically, this same dollar amount will be used as the disincentive amount in the contract. Engineering judgment or agency policy usually determines what percentage of the estimated RUC is used. The total incentive and disincentive amounts are calculated by multiplying the number of days that the contract is completed earlier or later than the contractor's bid time (B) by the established I/D amount for the contract.

The 1989 FHWA technical advisory proposes that the incentive daily rate should equal the disincentive daily rate. If different rates were selected, the incentive daily rate should not exceed the disincentive daily rate (FHWA 1989). A cap of 5 percent of the total contract amount is recommended as the maximum incentive payment. The 5 percent was based on a NEEP study of average incentive payments made on experimental I/D projects. The placement of a cap on the incentive payment limits the funding requirements that may result if the time analysis was not realistic for an accelerated project time. With experience, the agency may feel comfortable in not setting any maximum on the number of days for which an incentive can be earned. Frequently, the incentive is capped by the agency at a maximum percentage of the contract amount, a set maximum dollar amount, or a set number of days that will be paid for early completion. According to the FHWA Technical Advisory, a cap should not be placed on the maximum disincentive amount, although some agencies cap both the incentive and disincentive equally (Anderson and Russell 2001). Overestimation of the maximum incentive may waste public money while underestimation will reduce the effectiveness of the incentive to motivate the contractor (Shr and Chen 2004).

Jaraiedi (1995) developed a formula for evaluating daily and maximum incentive amounts. In the formula shown in Equation 2, on the left-hand and lower bound of the equation, both C_0 and C_d represents the contractor's cost of acceleration—C_0 is the one-time mobilization cost and C_d is the daily additional cost—ΔE_t is the targeted time of acceleration estimated by the agency; in the middle is the amount to be determined, I, which represents the I/D per diem; and the upper bound of the equation is the public cost of construction—represented by R, the total road user cost for the accelerated period and p, some percentage of 100% or less that the SHA is willing to spend toward accelerating the work. Note that it would be economically irrational for p to exceed 100%, although Jaraeidi proposes that in some unique circumstances (such as under extreme lack of safety), SHAs may decide to exceed the 100% figure and pay more incentive than the RUCs calculated using traditional means.

The Jaraiedi algorithm is useful to understand that I/Ds must fall between the contractor's acceleration costs, on the lower bound, and the public's road user costs on the upper bounds. Detailed procedures are not developed, however, for the means to set the lower bounds, nor to establish what the proper percentage of road user cost should be set for the upper bound. The determination of these values is left to judgment.

$$\text{I/D Amount Algorithm: } \frac{C_0}{\Delta E_t} + C_d \leq I \leq pR \qquad (2)$$

Shr and Chen (2004) developed a method for setting the maximum incentive for I/D projects. Using historical data from Florida DOT projects, they evaluated the relationship between construction time and cost, and proposed that, given an incentive amount, a contractor may most efficiently utilize those funds to accelerate the work for a calculable amount of time. Due to the complexity of the calculations, the final equations are not reproduced here, but the reader is encouraged to explore the development of this formula in the original paper noted in the References. The basic assumption of these equations is that all projects share similar fundamental characteristics with the relationship between time and cost.

Jaraeidi (1995) advises that each project situation is unique, and a project-by-project analysis of acceleration costs may be appropriate. Regardless of the technique used, it would appear prudent to recognize that in most cases, only portions of projects are accelerated and whichever formula for establishing I/Ds is used, the economics surrounding the accelerated portion should be considered.

Whichever method is used to determine the I/D amount, the allocation for the potential award of the entire incentive amount should be factored into the agency's construction budget (Anderson and Russell 2001). Since such allocation of funds may require significant prior approval, and in some cases legislative action, many SHAs establish maximum incentive amounts. Further, as the use of I/D-type contracts becomes more widespread, tracking of the potential outlay of many concurrent incentives during a short time period would avoid unexpected outlays in forecast expenditures. In order for such budgeting to be effective, agency cost forecasting systems should be developed which address the variability of incentives both in terms of amount and timing.

Liquidated Damages. The differences between liquidated damages and incentive/ disincentive provisions were stated in the FHWA Technical Advisory document. In contrast to the liquidated damages policy, which has as a prime function the recovery of construction engineering (CE) and/or additional costs associated with the contractor's failure to complete the project on time, the I/D provision is intended to motivate the contractor so that work will be completed on or ahead of schedule. Typically liquidated damages provisions apply to all projects; however, I/D provisions apply only to special projects (FHWA 2005).

The use of I/D provisions does not preclude the use of liquidated damages (LDs). Liquidated damages are typically applied to the total project time, while I/D assessments are typically applied to one or more selected project segments. However, caution should be exercised when establishing liquidated damages to ensure that the LDs do not cover the same costs of delay that are included within the I/Ds, avoiding a double charge to the contractor. In some cases, liquidated damages have been assessed only after a capped disincentive amount has been reached (Anderson and Russell 2001).

I/D Specifications

After the key parameters have been defined, the specifications are developed. As an alternative contracting technique, specifications must adequately draw the bidders' attention to the unique aspects of this contracting method. Standard specifications are in place in many agencies, and a careful reading of established specifications should be made to ensure that conflicts are eliminated.

Considerations should be given to standard specification sections dealing with bid award, scheduling, contract time and adjustments, definitions, liquidated damages and others. New specifications must be included that describe the incentive/ disincentive program. Guideline specifications that interleave with the AASHTO *Guide Specifications for Highway Construction* (1998 Edition) were developed in NCHRP Report 451 – *Guidelines for Warranty, Multi-Parameter, and Best Value Contracting* (Anderson and Russell 2001); the reader is encouraged to seek these draft specifications when developing I/D specifications.

Project Development

All involved parties, including agency design personnel, consultant design personnel, and agency construction personnel, must bring their expectations into agreement early in the process. During the development of I/D projects, extra effort should be made to ensure that the design, specifications, schedule, etc., are compatible and appropriate for the project. A field change to correct plan errors, especially those related to the I/D phase work items, will be very costly in both time and money on an I/D project. The plans and specifications should indicate any unusual condition or any restriction under which the contractor may be required to work, such as prohibiting jack hammering or pile driving during the night due to noise problems, or work restrictions related to environmental issues (FHWA 1989).

During the pre-construction phase of the project, all parties (e.g., local officials, police, local traffic engineers, construction engineers, etc.) should be involved in the project development. A partnering process would help agency and contractor personnel agree on project objectives, critical success factors, and other areas that might result in problems (Anderson and Russell 2001).

Pre-design field reviews are essential since "as built" plans or old construction plans may not be reliable. Maintenance operations, utility work, modifications on adjacent properties, or field changes done during the original construction but not recorded on the plans may result in substantial differences between the as-built plans and field conditions (FHWA 2005). The clearance of utilities and rights of way is critical to avoid time delays and related contract claims.

The I/D type of contract (*I/D only*, or *A+B+I/D*) must be identified as the chosen contract method early in the project. Design considerations, selection of construction methods and techniques, and sequences of work are all interrelated and will be impacted by the *A+B+I/D* process. A constructability review process would help ensure that these interrelated areas are well planned.

Contract Procurement

Pre-bid conferences, conducted by the agency, should be held when beginning the implementation of I/D contracts; contractors who are used to bidding other forms of proposals may misunderstand the intent or mechanics of the bid process. However, as the bidding community gains experience, routine pre-bid conferences may not be required, unless the complexity of the project warrants a meeting (Anderson and Russell 2001) (FHWA 2005). The advertisement of bids for an I/D project generally follows the same process as used for traditional contracts, with the exception of noting the project as an I/D project.

For *A+B+ I/D*, public bid openings are generally the same as for traditional bid openings, except that both the cost (A) and time (B) bids as well as the total equivalent value of each bid proposal may be announced, or only the (A) portion is announced and the calculation of equivalent bids are made later, and the successful bidder choice is published separately.

Pre-construction conferences provide an important opportunity for the agency to highlight the reason for the I/D provision and to discuss in detail the segment which is the focus of the incentive. All parameters of the I/D provision should be discussed, including any potential interferences or concerns held by the contractor. The NCHRP report (Anderson and Russell 2001) recommends that discussion include: the awarding of additional contract time (if the contract provisions allow for time extensions); when time charges begin; when time charges end; and utility interfaces, if not cleared. Additional items may include the contractor's anticipated methods of acceleration and any expedited administrative procedures that may be required to avoid delay.

Contract Administration

During the construction phase, cooperation and coordination between the contractor and the agency are essential, since any delay in approval of change orders can be costly. Decision-making and approval must be promptly provided to the contractor at all times that I/D work is in progress. These discussions should consider future critical operations and potential problems. Communication among major involved parties is crucial throughout the duration of the project. Coordination between the parties may require more input from design personnel and the contracting community, and planning should reflect this additional effort. During the project, conflicts must be resolved in a timely manner to avoid delays.

Hopefully, any pre-existing conditions that affect the efficient prosecution of work were discovered in the pre-design effort (discussed under Project Development). Since the effects of such conditions may be substantial, agency project managers should verify that conditions such as existing utilities, environmental contaminations, rights-of-way, local regulations concerning noise abatement, etc. have been properly anticipated and addressed.

To facilitate the project engineer's ability to make prompt decisions, the contractor should be required to submit a CPM schedule for review and approval prior to starting work. In addition, since the schedule will be used to gauge and analyze the contractor's progress, determine time adjustments, and evaluate claims, the contractor should be required to update the CPM on a regular basis, which might be in conjunction with the regularly scheduled job site progress meetings. Regular meetings to update the CPM serve as a valuable contract administration tool, especially if any changes occur. During the life of the contract, the contractor must meet all milestones and completion dates.

Especially critical is the handling of claims for time extensions. Due to the increased sensitivity and cost of time changes (caused by the assessment of I/Ds), time extensions receive significant scrutiny. Therefore, potential delays should be anticipated far in advance and avoided wherever possible. The *A+B+I/D* method of contracting assumes that the contractor will apply whatever means are available to meeting and/or accelerating the incentivized segment; requests for extension should be viewed against this standard. Project meetings with the contractor regarding avoidance of any potential delay should be held regularly and in advance of construction of the incentivized segment (FHWA 1989; FHWA 2005). Additional work should be anticipated, but should be limited to that which does not affect the critical path of the incentivized segment.

Physical construction is usually performed in the same manner as in traditional competitively bid contracts. However, many *A+B+I/D* projects use extended working hours, either with overtime or multiple shifts, so that the contractor can obtain the maximum incentive possible. The agency must be ready to adapt its work schedule to the contractor's work schedule. This often involves additional inspectors to verify the contractor's work and faster turnaround times on testing. Agency personnel must make sure that they do not delay the contractor and should emphasize timely decision making.

While the total number of agency personnel hours may not necessarily increase for such activities as quality control inspections, the volume of inspections over time may increase due to an accelerated schedule. This may require additional inspectors and many agencies may have insufficient personnel to cover such temporary needs. In response, agencies may react through shifting more quality control responsibility to the contractor.

Support from agency upper management and industry as well as supervisory training are critical success factors for I/D contracting. Implementation of I/D contracting will require agency personnel with appropriate levels of skill and expertise. NCHRP Report 451 (2001) notes that "agency personnel must be trained to respond quickly to contractor RFIs and potential change orders to eliminate contractor delay."

Evaluation of Project Effectiveness and Feedback Process

The final step of the I/D contracting process presented in Figure 1 is the evaluation of the project regarding the effectiveness of the I/D contracting method. The decision of the agency to continue using this method for similar projects should be influenced by this evaluation and any recommended changes for refining the I/D process should be implemented.

NCHRP identified several important items for effective evaluation in its guidelines for multi-parameter contracting, summarized below:

- Time: comparison among the engineer's time estimate, time bid, and actual project duration;
- Resource requirements: overtime hours, total hours and number of personnel required;
- Design process: track number of requests for information (RFIs), change orders, design hours, special reviews (constructibility reviews, etc.);
- Quality of the project: track performance of the project product over future life;
- Claims and litigation: number and value of time-related project claims, value of litigation, if any;
- Project costs: compare engineer's cost estimate, price bid, and actual project cost, I/D earned;
- Innovative construction processes: the processes used by the contractor and/or subcontractors to speed up construction should be noted;
- Innovative sequencing used: traffic or construction sequencing used by designer or contractor;
- Impacts of third party conflicts (if any): measure delays and their cost to agency and contractor; and
- Overall effectiveness of the project (Anderson and Russell 2001).

Feedback of lessons learned in the project is essential and should be requested from all project parties, including contractors, the agency, FHWA personnel and others. The agency has to make a policy decision whether they should continue further application of I/D contracting according to the selected method for similar projects, or revise methods or contract techniques to expedite future projects; comparison of project results against initial project objectives will provide guidance.

Future projects should be evaluated for indication of needed improvements (Anderson and Russell 2001). Improvement processes may include recommendation of necessary changes to: specifications, design process, revision of the project selection criteria, revision of the risk management process for agency resources and risks, and revision of the I/D time and I/D amount parameters. A feedback loop should be created to allow lessons learned during execution of the project to be embedded into future design and project procurement documents.

Conclusion

This research has consolidated diverse guidelines for implementing an Incentive/Disincentive contracting process to be used by State Highway Agencies (SHAs) and others, and has proposed a holistic methodology for developing I/D projects. The implementation process includes the identification of goals and needs for accelerating a project schedule, evaluation of those projects with I/D criteria, selection of type of contract, incentive type, implementation of a risk management process, determination of key parameters, preparation of specifications, contract administration and evaluation process.

Future Research

Future development of this process will continue as more experience with I/D projects is gained, and as more data is collected and indicated changes are implemented. Particularly important in advancing the use of I/D provisions will be the discovery of methods to evaluate the setting of I/D amounts based on bidders' perspectives, especially the lower bounds—important in ensuring that the use of I/D obtains the most effective use of public funds. Development of the upper bounds of I/D amounts—Road User Costs—are well-developed; yet, whether a particular I/D amount acts as a true contractor incentive on a given project is yet to be fully understood.

Improved methods for creating realistic acceleration schedules for project segments is a prerequisite for accurately forecasting contractors' acceleration costs—key to establishing fiscally-responsible I/Ds. SHAs have not historically tracked production rates in a manner that provides sufficient information for such analyses. Discovery of methods to accurately predict realistic durations under a variety of acceleration scenarios would add more certainty to the application of I/Ds.

As A+B+I/D contracting becomes more prevalent and a "portfolio" of I/D projects aggregates, additional problems arise that should be monitored and exposed through research. Two situations must be understood. First, the problem arises in which a contractor that has several current projects may purposefully shift resources and delay one project in favor of another project on which the gain of an incentive is more beneficial. Second, techniques should be discovered to deal with the variable nature of incentive expenditures on a portfolio of I/D projects. Since these expenditures may not be included in base project forecasts, the concurrence of many incentives during a short time period may create unexpected drains on short-term agency budgets.

Finally, as discussed in this paper, I/D projects require revised methods of administrative and project management personnel. It is unclear as to whether this places a substantial burden on the agency and as to whether there are any significant short- or long-term effects on costs, the quality of the product, or of the quality of the design or construction phase management. Research to baseline and evaluate any such variant costs would benefit agencies' abilities to properly plan for the real cost of such projects.

Acknowledgment

The authors would like to express thanks for the generous support of the Oregon Department of Transportation (ODOT) and the Federal Highway Administration (FHWA); the development of this paper is a portion of a larger research project into the use of incentives and disincentives in transportation projects, funded by ODOT and FHWA. Particular appreciation goes to the members of the Technical Advisory Committee and additional key participants— Bob Callaway, Dale Deatherage, John Gernhauser, Jeff Graham, Tom Lauer, Felix Martinez, John Reidl, and Roberta Watson.

Notation

The following symbols are used in this paper:

A = Proposed cost bid for completion of a project.

B = Proposed time bid for completion of a project, in days.

C_0 = The one-time mobilization cost incurred by a contractor to accelerate a designated segment.

C_d = The daily additional cost incurred by a contractor to accelerate a designated segment.

E = Time estimated by agency for completion of a project or designated segment, in days.

I = The I/D per diem amount.

P = Some percentage of 100% or less that the SHA is willing to spend toward accelerating the work.

R = The total road user cost for the targeted accelerated period.

s = incentive/ disincentive daily rate.

x = Actual time of completion of a project or designated segment, in days.

ΔB = B minus x, the time variation between bid and actual completion times.

ΔE = E minus x, the time variation between agency-estimated and actual completion, in days.

ΔE_t = The targeted time of acceleration for a designated segment., estimated by the agency, in days

References

Anderson, S. D., and Russell, J. S. (2001). "Guidelines for Warranty, Multi-Parameter, and Best Value Contracting." *NCHRP Reports*, <http://gulliver.trb.org/publications/nchrp/nchrp_rpt_451-a.pdf > (August, 2005).

Arditi, D., Khisty, J., and Yasamis, F. (1997). "Incentive/Disincentive Provisions in Highway Contracts." *Journal of Construction Engineering and Management*, 123(3).

FHWA. (1989). "Incentive/Disincentive (I/D) for Early Completion - T 5080.10." *TECHNICAL ADVISORY*, <http://www.fhwa.dot.gov/legsregs/directives/techadvs/t508010.htm > (August, 2005).

FHWA. (1991). "Report on the 1990 European Asphalt Study Tour." *International Technology Exchange Program/ Infrastructure/ General*, <http://international.fhwa.dot.gov/Pdfs/Eurotour.pdf > (August, 2005).

FHWA. (1994). "FHWA Summary Report of the Contract Administration Techniques for Quality Enhancement Study Tour (CATQEST)." <http://ntl.bts.gov/DOCS/catqest.html > (August, 2005).

FHWA. (2002). "Design-Build Contracting." D.of Transportation, ed., National Archives and Records Administration.

FHWA. (2005). "Construction Management Practices in Canada and Europe." *International Technology Exchange program*, <http://www.international.fhwa.dot.gov >

FHWA. (2005). "Contract Administration Core Curriculum - Participant's Manual and Reference Guide 2005." *FHWA > Infrastructure > Program Administration > Contracts*, <http://www.fhwa.dot.gov/programadmin/contracts/coretoc.htm > (August, 2005).

Herbsman, Z., and Glagola, C. (1995). "Lane Rental - Innovative way to reduce road construction time." *Journal of Construction Engineering and Management*, 124(5).

Herbsman, Z. J., Chen, W. T., and Epstein, W. C. (1995). "Time is Money: Innovative Contracting Methods in Highway Construction." *Journal of Construction Engineering and Management*, 121(3), 273-281.

Jaraiedi, M., Plummer, R. W., and Aber, M. S. (1995). "Incentive/ Disincentive Guidelines for Highway Construction Contracts." *Journal of Construction Engineering and Management*, 121(1), 112-120.

Lee, E. B., Harvey, J., and Ibbs, C. W. (2004). "A Computer Simulation Model: Construction Analysis for Pavement Rehabilitation Strategies (CA4PRS)." <http://www.ce.berkeley.edu/~eblee/CA4PRSModel_Tech_021204.pdf > (October 25, 2005).

Shr, J.F., and Chen, W. T. (2004). "Setting Maximum Incentive for Incentive/Disincentive Contracts for Highway Projects." *Journal of Construction Engineering and Management*, 130(1), 84-93.

Transportation Reserach Board. (1991). "Innovative Contracting Practices." National Research Council, Washington, DC.
Utah Local Technical Assistance Program (LTAP). (2005). "A+B Bidding." *INNOVATIVE CONTRACTING*, <http://www.ic.usu.edu/ > (August, 2005).

Van Ness, N. J. (1990). "Memorandum, Innovative Contracting Practices Initiative and Special Experimental Project No. 14." *Infrastructure/ Program Administration/ Contracts*, <http://www.fhwa.dot.gov/programadmin/contracts/021390.htm > (August, 2005).

Weseman, W. A. (1995). "Memorandum, Special Experimental Project No. 14." *Infrastructure/ Program Administration/ Contracts*, <http://www.fhwa.dot.gov/programadmin/contracts/050495.htm > (August, 2005).

Index

A + B contracting 81, 82–84 , 132, 138–139; schedules 84
average daily traffic 86

bidding process 61; award criteria 70–72, 71*t*; bid preparation 12; competition 5; documents 122; documents, review of 63, 92, 93*t*; literature review of 62–63; prequalification phase 9, 11–12, 99, 147; prequalification phase, advantages and disadvantages 100–101; proposal phase 10
bonds. *See* performance bonds; surety bonds; warranty bonds

case reports 64*t*, 77–79*t*, 126–127; Central Texas Turnpike 6–9
Central Texas Turnpike 7*f*, 8*t*; case report 6–9
certification 102
commercial development plans 54–55
communication 6, 125
Comprehensive Development Agreement (CDA) 1, 10*f*; benefits of 4–5
construction costs 85
construction jobs 47
construction manager, role of 52–53; selection of 54
Construction Manager-at-Risk project delivery method 50–51; benefits of 56–57, 58; Design-Bid-Build method, compared 56; organization chart 55*f*; origin of 52
construction methods 125; selecting 122
construction projects, itemized 53
contract administration 99, 147–148
contract management 13; best-value concepts 63, 64*f*
contracting 3; non-traditional 4
contracting techniques 114, 132; evaluating 81, 87–91; guidelines 132; selecting 137–140; studies 131
contractor liability 22; contract clauses 43–44, 45
contractor qualifications 98, 109*f*; rating systems 108–109; third party evaluation 102, 103*t*
cost plus time contracting. *See* A + B contracting
costs 65, 123, 123*t*; administrative 85, 93–94, 96; incentive/disincentive contracts 143; road user 86, 94–95

defects, defining 27–28
design 125
design changes 66, 148
Design-Bid-Build method, Con struction Manager-at-Risk project delivery method, compared 56
design-build delivery method 1–2, 81–82; guideline documents 5; study recommendations 5–6, 9, 15–16
dispute resolutions, contract clauses 42–43

economic impact 47–48
emergency conditions, contract clauses 41
European contract practices 130–131

fast-track construction 52
federal funding 47, 48
Florida Department of Transportation 47

funding 48, 49*t*, 57–58; allocation 145

incentive/disincentive contracts 81, 82, 138; completion time 141–143; implementation 133–134, 133*f*; initial uses 130; project selection criteria 136–137; specifications 146; types of, comparison 139*t*
incentives, determining amounts 144–145
inspections, schedule 22–23; contract clauses 39–40
International Standards Organization (ISO) 102

lane rental contracting 81, 82, 132, 140
legislation, public construction projects 61–62
liability 140
licensing 100
liquidated damages, incentive/disincentive contracts 145
loss of service 81, 86, 132

maintenance 23–24, 29; contract clauses 44
Miami International Airport, traffic congestion 46–47

NCHRP guidelines 132–133

pavement performance 127
performance bonds 99, 101
performance criteria 21–22, 85–86; asset items 21–22, 32; asset items, deterioration of 28–29; contract clauses 38–39, 39*t*; defining 27, 28*t*; rating 23*f*, 24*f*
performance-based contract 21
personnel recruitment 9, 65, 98
pilot projects 119, 121
planning process 6
project administration 122
project control 22
project delivery methods 3, 19*t*; best value parameters 65–66; comparison 50–51, 51*t*
project duration 141–143
project evaluation 6, 55–57, 66–68, 67*t*, 106*t*, 109–110, 122–123; incentive/disincentive contracts 148–149 ; methods 105–106; models 103; quality assurance tests 108; questionnaires 107; rating systems 68–70, 69*t*, 70*f*, 87–91
project financing 8
project management 13, 85, 146
project planning time 85
project selection 126, 135–136
project team 9, 52

quality control 14, 53, 65–66, 86
quality performance measurements 101–102

railroads, co-location with highways 9
regulation, public construction projects 61–62
remedial work, contract clauses 40–41
rental car facility 49–50
request for proposal 10
research needs, contract techniques 150
right of way 14, 49
risk management 141

scheduling 65, 125, 134–135, 141
Special Experiment Project Number 14 80, 132
state departments of transportation 3; contract clauses 44–45; contract preferences 85; management goals 113–114, 125; needs and objectives 117; specifications 120*t*, 121*t*; standard specifications 119

state legislation, public construction projects 61–62
surety bonds, contract clauses 37–38
surveys, questionnaires 63, 72–73, 82, 86–87, 103

Texas Department of Transportation 1–2
toll roads 6–9
training 5–6, 124
Transportation Equity Act for the 21st Century 48
transportation modes, multiple 50

utilities 14

Virginia, Public-Private Transportation Act (PPTA) 20

warranties, coverage, contract clauses 38–39; documentation 37–45; highway construction 26–27; use of 117
warranty bonds, contract clauses 37–38
warranty clause projects 81, 132; appropriateness 125; costs 123, 123*t*; guidelines, lack of 114; implementation 116*f*, 123–124; models 114; objectives 115; participants 117–119; review by state agencies 115
warranty contract clauses 22, 24–25, 25*f*, 31*f*; advantages and disadvantages 33; asset items 32; length of warranty period 28–31, 30*f*; research methodology 25–27
Wisconsin Department of Transportation, case report 126–127